S0-DNK-920

WEALTH OF NATIONS IN CRISIS

by

RONALD C. NAIRN

"... But man, proud man!
Dress'd in a little brief authority,
Most ignorant of what he's most assur'd,
His glassy essence, like an angry ape,
Plays such fantastic tricks before high heaven
As make the Angels weep...."

Measure for Measure
— Wm. Shakespeare

Library of Congress Catalogue Card Number 79-90284

ISBN 0-934-018-00-6

Printed in the United States of America.

First Edition

TO MARY

ACKNOWLEDGEMENTS

I am deeply indebted to the Woodrow Wilson International Center for Scholars, Smithsonian Institution, Washington, D. C., for the opportunity accorded me by them to write this book. In particular I thank the Board of Trustees, Dr. James H. Billington, the Director of the Center and a very fine staff, each of whom was unfailing in counsel, courtesy and practical assistance.

I wish also to thank three colleagues at the Center for special help: Father Joseph Sebes, S.J., Georgetown University, Washington, D. C.; Dr. Moshe Lewin, University of Birmingham, United Kingdom; and Dr. Chiaki Nishiyama, Rikkyo University, Japan. I offer special thanks to Dr. Nishiyama, a pioneer in the concept of human capital.

As I have so often in the past I turned, too, to a long standing mentor and friend, Dr. David Nelson Rowe, Professor Emeritus, Yale University who read this manuscript and gave invaluable advice.

Particular thanks are also due to Mr. Tom O'Keefe who patiently and expertly proof-read this volume.

I was also fortunate to have the assistance of many fine institutions among them the Library of Congress, National Agricultural Library of the U. S. Department of Agriculture and the National Aeronautics and Space Agency.

My special thanks, however, is to my associates in my former company, Rubel, Bates, Nairn and Co., Phoenix, Arizona. To Jack Rubel, to C.T.R. Bates, to Marvin Gibson and Copeland Howe — not only thanks. I owe special thanks to my present co-worker Irvin E. Spitler, who personifies the person who translates ideas into productive actions.

Special thanks is also due Mrs. Eloise S. Doane, Woodrow Wilson International Center for Scholars, who supervised manuscript typing.

Then there are others who can be thanked only generally. These are countless farming folk all over the world whom good fortune has allowed me to know and work with. If my words can help, even in some tiny way, to make their lot better, that will be reward enough.

I wish to acknowledge with thanks the support of the Earhart Foundation for research (formerly the Relm Foundation) in Southeast Asia, much of the results of which have found their way into this study.

CONTENTS

PREFACE

This is a book about human capabilities and how, as a phenomenon of our age, those capabilities are being thwarted if not crushed. As we shall see the price we pay for this denigration is awesome.

Alternatively some might say the book is about economics or politics. This point could also have validity. But we often forget that economics and politics are but reflections of human action.

The contention is that humans are the alpha and omega of all that happens to this planet. This truism is now subordinated to the phenomenon of our age, a reversal of role. We have, since about the French Revolution, assumed that institutions are more important than people, especially the institutions of the State. This is pure illusion and daily the evidence mounts to prove the point.

This book is developed around human productivity in its most basic form. Why it is that our planet, endowed as it is to produce stupendous wealth, does not meet the most elementary needs — sufficient food to nourish each individual's body and sufficient fiber to clothe and shelter him adequately.

The issue is the abolition of poverty. This begins with productive agriculture.

If a nation mainly comprises farm folk, as most nations still do, economic development begins only through optimizing the cash flows of the bulk of the nation's people — that is the farming folk. Short of a nation possessing some modern version of King Solomon's mine, such as vast oil reserves, economic history tells us this is the way economic development begins. For modern instances we look now at South Korea, Japan after Meiji and especially after its destruction in World War II; at the economic miracle of the Republic of China on Taiwan. Economic development began with the farmer — the individual farmer.

As far as a nation's internal economy is concerned the economic spillover effect of increasing incomes of the 70 or 80% of the population who are farmers, ofttimes by many hundreds percent, is as obvious as it is spectacular. So too could be the international effect. Of the four billion plus humans now inhabiting this planet some three quarters of them approximately are small peasant farmers. By the year 2000 the planetary population will be about six billion. An even greater proportion of these than is now the case promise to be peasants. As of now the majority of peasant incomes on the planet hover at about the subsistence level. Nevertheless, as one example, the USA has a trade flow of about $30 billion per annum with this group. This is about the same as US trade with Europe. But Europe is a mere 250 million people — not three or more billions. Europe's economy is fairly well optimized and trade expansion is likely to be in small increments. The opposite applies in the peasant world. Their income optimization can be massive and is spread over vast numbers. Instead of an annual trade of $30 billion with the peasant world the U.S. should be contemplating ten times that amount — but only if peasant agriculture as a first step attains its enormous potential for high productivity. These are the kinds of material stakes we are playing with.

Why does this not happen? Or, why has it happened in certain countries and not at all in others? As we shall see we have the material resource and we have the human resource. What then inhibits?

The inhibitors are in location, powerful, pervasive, extensive and not ready to give way easily. The core of the inhibition is man himself. On one side stands man with his richness of potential, his imagination, his creativity, his skill and energy. On the other side standing ever brooding and omnipresent is the Manichean devil — other men always abusing their power and license over their fellows. It is not the resource potential of the planet where our shortfall arises. It arises in the minds and in the related actions of men as regards to other men.

The forms of the Manichean devil are easy to define. There is the overweening desire of some men to dominate the territorial integrity of others. Thus in the international field we see a planet torn by wars and related tensions. This does not mix well with good farming. Then there are the ideologues who assume the extraordinary role of becoming creators of "new men" and who may use agriculture as a bludgeon in this process. There are politicians and their attendant bureaucracies seeking short term interest and applying palliatives conceived in ignorance. Agriculture, which is nature, will not tolerate shortsightedness, capricious decisions, irresponsible demand or neglect. Last comes culture. It stands by itself. But because it determines the way we humans view our world and ourselves in it, no examination of agriculture can deny its presence.

Waiting in the wings is yet another potential inhibitor. This is the environmental movement, which especially in its "no growth" configuration belongs to the affluent West for no one else can afford it. These cruel people behave like a third generation inheritor of great wealth. They care more about their fish and wildfowl than they do about the impoverished mass outside the estates' walls. "No growth" condemns billions to poverty in perpetuity.

The creation of wealth, which is another term for skillfully performed human work, is the creation of individuals. Imagination, intelligence, creativity, skill, energy are not institutional attributes but human attributes. Institutions, particularly governments, do not create wealth but they can help creation by providing environments wherein the individual has liberty to optimize his talent. Institutions, especially governments, must be servants of man in this process. Instead they have become master and often destroyer.

In a sense this book is a parable and the parable relates to all productivity. Agriculture is the focus because it is basic to all other forms of productivity. But we, the people of the 20th Century, are all in the grip of irrational forces. These, the inhibitors, inhibit all productivity because these irrational forces impact on all people: be it carpenter, banker, machine operator or plant foreman, shopkeeper or artist, professional or student, the newborn, the old and especially our peasant — the inhibitors inhibit all. The fundamental question in the parable is an old one. Does the State exist to serve the individual or the individual the State? The irrationality of the 20th Century is the State has superceded the individual when the State needs the individual most. The cry of the world's poor is for alleviation of their poverty. Only individuals in a condition of liberty can effectively produce wealth and then go on to optimize that effectiveness again and again. Each State as master has by definition assumed responsibility for answering the cry of its poor. And the State as master is failing for the poor remain poor when they might have affluence. The task of the 1980's is to call the State to account: to remove those institutions of the State that are the primary enemy of the wealth of a nation.

FUTURE 1

To meet the future we must create a new positive state of mind toward agricultural development: a belief that the planet is a potential cornucopia, and that humans' innate capacity to handle the agricultural problem belongs to all of us. There is no single best way to select means to solve the problem. There are only parochial, localized best ways. This means devising a "mix" of traditional and modern usages that fit the economic and human aspects of particular plants, animals, soils, climates, and people at particular times and in particular places.

Everywhere in the developing world peoples *are* innately capable of meeting this problem, but almost everywhere they are prevented from doing so by ideologies, bureaucracies and politicians. Under current ideologies, bureaucracies and politicians tend to be masters, not servants of people. This must be reversed. The call is for individual decision-making, not only to enhance production but for the personal satisfaction which autonomous decision-making can bring, for increased production — even satiation with goods — is not enough to bring human satisfaction and spiritual well-being.

A general return to limited government, with the state as servant, not master, would enhance the power and influence of politicians, and not — as so many believe — weaken them. This is because too many politicians today undervalue human beings who are, in fact, the primary resource of the state and economy. From this undervaluation flows the tendency of many in government to believe that human populations are a curse and a burden to development. So they try to buy off the poor with welfare instead of developing them as the prime capital of society. In turn, these politicians and bureaucrats are disliked and discredited by their human constituencies, creating a vicious circle of misuse and mistrust. This, in turn, inhibits creativity.

As it is not under totalitarianism, human creativity under freedom is a low-cost and self-sustaining resource. But freedom is in short supply. The urgent need is for re-politicizing toward freedom. Much of our current ideological jargon must be jettisoned, along with most of the so-called "social programs" which are designed heavily for the supposed advantage of politicians and bureaucrats. The idea that government must serve *in loco parentis* is an assumption that flies in the face of what people can really achieve under control of their own affairs. As we shall note later, Russian and Chinese peasants have demonstrated this by differences in production between collective and state farms and their private plots of ground. This, too, we shall examine.

Until we regain control of our own affairs through dissolution of the centralist power of government, we cannot regain individual responsibility and take action, autonomously, to obtain the best results from human work. Is so-called "modern society" too complex for this? It is indeed, but such societies need not exist. They must, in fact, undergo evolution toward human freedom. Without this, low-level poverty will persist. The West will continue to try to alleviate it by financial aid, hoping the poor will simply be bought off. Along with this will be renewed many of the current demands for a "new order" of economics, i.e., intervention to destroy the market system and all other human freedoms with it.

The West, instead, should be making counter-demands of developing areas. As the price of any real help it can give to agricultural development, it should insist upon basic administrative and political reforms in the Third World, moving away from those all-too-prevalent tamperings with economics. For example, there is the prevailing practice of internal price-fixing, the attempt to abrogate normal supply and demand in the market. In Argentina, under Peronist policies, for example, the government tried to keep prices low for its chief constituents by forcing farmers to sell their wheat at 30% world market price, and beef at 40%. Such policies were combined with massive government deficit spending for other purposes. The result was supposed to keep general price levels down. In actuality, in 1976, the inflation rate was 444%.

This could only devolve into political turmoil. Over the past decade political assassinations in Argentina have killed more people than in the entire French Revolution. All this can be repaired eventually, but only by creation of a new political order. Can the West throw off its illusions and see the truth? It is to be feared that the West has created its own myths and has lost the ability to see things as they really are. It is incapable of making upon the Third World the courteous but firm demands that are required for the welfare of all. From this, the West itself will suffer. Perhaps this will only increase the guilt feelings of the West, which will probably tend to blame itself for world suffering. Such situations simply invite international intervention by troublemakers, and there are plenty of those around.

It is this writer's prediction, born of some thirty years of association with the underdeveloped world, that if the truth is told — courteously but accurately — problems may not quickly disappear, but at least the Third World will show a new respect for the West. From this some constructive dialogue might emerge. This would be a notable and essential first step.

What we really seek is to recapture an ethic...a personal ethic which assigns a focal point to the individual, gives him a rationale for his own development and therefore his society's development. So much of what is being taught, whether it be politics, economics or agriculture, is without any reference to the nature and purpose of man. Yet man is the reason for the entire exercise. We have tended, however, towards an inversion. We have created constructs and we try to fit men into them. Instead, we need to construct a metaphysical exploration of man as producer and consumer because, in producing and consuming, man also has end and purpose. We must have purpose — purpose as basic as providing food and shelter or as metaphysical as a search for ultimates for man. In the end, as it ever does, the practical problem becomes an ethical problem.

We can prescribe for the world's farmers an ethos in seven parts. The components are drawn from this book. Without them it is hard to envisage any significant agricultural development occurring in the future.

First and foremost is that each individual farmer have security of tenure on field and home. This does not necessarily mean ownership of land. It is simply good common sense to place stewardship of a specific area of productive land in the hands of those who work the area and to insure that stewardship is perpetuated. If one believes in human capital, what better way to insure its agricultural flowering than through assigning a particular piece of farmland to an individual so that he can particularly and intimately superintend the husbandry of that land? The alternative, as has been demonstrated on a grand scale in this century, is for the state to subsume this individual superintendence.

Here we have the perfect dichotomy of superintendence — on one hand the amorphous, generalized, remote and ignorant state; on the other, the definitive, specific local individual. Improved husbandry can hardly be achieved other than by individual superintendence. If, however, the argument is for some ideological or political construct removed from agriculture, the state then has its amorphous way. We should be clear in our minds as to the objective being pursued. Is it an agricultural goal? Or an ideological or political one? There are differences, yet often the discrimination is not made.

The second factor in the ethical prescription is freedom to make decisions affecting the general welfare of field and family, paying homage to an old simplicity that "the man on the spot knows best." Increasingly, however, the powerful centralized state subsumes this notion: it is the state that knows best. Of course the state does not know much at all. Nevertheless, the state has a kind of misguided prestige to uphold; in all its majesty it must proclaim that it surely knows what is good for specific parts of the commonweal.

This notion has probably created as much mischief as any single feature of contemporary life. If farming is to be more productive, we have to reverse the trend. It is suggested that this is as much in the interest of the state as it is of the individual. Increasingly, the state appears as a great ass. When it pursues operational roles it blunders, stumbles, harms and hurts. Increasingly, especially in the advanced democracies, it becomes an object of ridicule. Under these circumstances the state tends to become even more arbitrary. Thus it accentuates its own follies and idiocies.

The state could save itself from these peradventures by relying upon individuals. The man on the spot is not always right; he is simply apt to be more right more often. Moreover, in reality, there are many men on the spot, each making decisions. All will not be wrong — there will be a majority who are right. Hence, progress is possible. The state can even be saved.

The third element in this new ethic is that the individual farmer must have freedom from penury-inducing taxation and other levies. Again we ask for a reversal of a trend. All over the world we have witnessed the growth of centralized states, and, at the same time, an awesome growth in state budgets. It is nonsense to say that these staggering budgets are spent "for the good of the people." When the state regards poor people as a liability rather than potential assets, it cannot eradicate poverty. Thus its remedy is to buy off the poor. For this increasingly large sums of money are required.

Governments do not produce wealth. For a government to have funds to expend, it must first extract those funds from its people. The state does this with increasing recklessness, if not insolence.

Another trend must also be noted: governments penalize success, penalize those who more effectively produce wealth. Progressive taxation is the name of this process. If current trends are maintained we should expect more — not less — taxation of successful farmers.

We should never forget the significance — totally underestimated, it seems — of envy and spite in politics. It arises from success in the production of wealth. The state will manifest this in many forms: in taxes and other financial and regulatory penalties.

What is our individual farmer to do in his defense? He must band together with his fellows and demand limited government. This is a far from easy task, especially when his primary role in life is to be a farmer. Governments seem reluctant to limit themselves, even though powerful arguments can be made that it would be in their long-term interest to do just that. Governments have to be limited by people.

It can be stated categorically that there will be no agricultural development without freedom from penurious taxation and other levies. Human capital will not develop if it is penalized for its successes. Yet government, rapacious for wealth to feed its own desires, plans and interests, categorically denies this truism.

The fourth element in the ethic is that farmers must have access to credit at market rates. Credit is the lifeblood of agricultural development. Banking services are needed to lend money equitably to individuals at the lowest rates the market can sustain — again, an exercise in human development. As we shall note, loan officers are needed in every rural area in the Third World. By definition, a loan officer needs some education and training in the calling. This means that loan officers need to be drawn from the city, and so we come back to that other problem facing virtually all of the Third World.

Most young city-based Third World elites despise the countryside, especially the notion that they might have to reside there. Yet their services are essential if rural credit is to be available, perceptive and fair. The state needs to provide the environment, the most potent factor of which is freedom. In this case, freedom for banking systems to arise and grow in rural areas which, in turn, give adequate rewards to their servants. The task is not insurmountable. Some countries have developed successful rural programs in health care and in education. The same is now needed for finance.

A fifth element in the ethos is freedom for the individual farmer to sell his product to whomsoever he pleases. For some reason when this notion is put to Third World governments it displeases them most of all. One senses that they see freedom of an open market as a distinct threat to government control of individual lives, a weakening of governmental superstructure. This is often the very opposite from that which they are trying to achieve. It strikes at the idea of indirect taxation through import or export taxes. It strikes at state control of finances. It weakens the concept of the state as omnipotent. Instead, it builds up the idea of the individual as sovereign and this does not always sit easily with politicians and bureaucrats, their rhetoric to the contrary.

Nevertheless, some of the best returns for Third World agriculture have been accomplished by farmers dealing directly with processors. The most notable recent examples have been The Republic of China on Taiwan, where Farmers Associations have negotiated long-term contracts with U. S. and other foreign multinational corporations. It can be taken as a principle that individual growers will negotiate a better contract than will the state on their behalf. Such contracts are likely to be more financially viable, more realistic in terms of product quality, quantities and delivery times, and more flexible in terms of unforeseen changes. One has only to experience the two kinds of negotiations, private as opposed to government, to grasp the point immediately. Yet government is often extraordinarily reluctant to pass this freedom on to individual farmers. Without it, however, agriculture is unlikely to progress.

The sixth condition is actually an amalgam of the previous five: freedom for the farmer and his family to enjoy the fruits of their labors. It is so obvious. Why mention it? Because of the prevailing tendency, most manifest in the Peoples Republic of China and especially in the Soviet Union, where the farmer has been made a wage earner. In a sense, a peasant bound to traditional agriculture also has many of the features of a wage earner. In traditional agriculture there is a striking similarity of income level along with a limited chance to change income levels. The traditional agriculturist often does have other freedoms that state and collective farm workers do not, but the similarity of economic status should not be overlooked.

By paying farmers wages, conditions postulated in the first five freedoms are all negated. As a wage earner the farmer has no interest in security of field; he does not need to make important decisions regarding general welfare of his field; he is disinterested in taxes except insofar as they affect his wages; rural credit is an irrelevancy, as is freedom to produce and sell to whomsoever he chooses. He enjoys not what he produces, but what he is paid. Here, then, at one fell swoop the state can redress weakening of its powers. Instead of autonomous individuals responsible for a given piece of land, let the state be responsible and pay the individual a wage. Governments being what they are, it is probable that making farmers wage earners as in the U. S. S. R. is an idea whose time has yet to come. It bodes ill for agriculture's future and the only defense is a demand for the opposite: freedom to enjoy what one produces.

The seventh and last element in the ethos is communication, if not consultation. What are the goals? Who sets them and upon what basis? Are they achievable? Governments have preempted the sponsorship role in development. Even where they are inactive they normally will not let others substitute for them. Thus government must set goals, even establish blueprints. It will fail on both counts (let alone in operation) if it does not then go further and consult with those who will do the work.

Here I am talking only of sponsorship for development from traditional to modern. There was no need for the government of the Republic of China on Taiwan, for example, to set goals for its developed peasant farmers in negotiating contracts with overseas and in-country food processors. The farmers could do this themselves, as have other farmers in other parts of the world. In the development stage, the individual farmer represents a potential resource. He must be consulted within terms of operational capacities, especially on issues of motivation and incentive.

An ethos, however, is only a theoretical starting point. Agriculture is a practical affair. It demands that the ethos move from words into practice. The ethos must relate in practical ways to human needs, human capability, human creativity and human aspiration if it is to be worthy of human consideration. Against this strict criterion it may then be asked: is it possible? Can underdeveloped nations realize the ethos? An answer, full of hope, is that some already have.

In Japan, especially after World War II when Japan was certainly an underdeveloped nation in the material sense, we see the ethos in action. It produced spectacular agricultural results. More significantly, if Japan so wishes, it has human potential to make further spectacular gains in yields even from its present high base. This potential arises while much of world agriculture stagnates. Some may argue that it required the MacArthur "Shogunate" to create this situation, as did a similar authoritarian rule after Meiji. However valid this commentary, without doubt the freedoms espoused by the ethos are today real in Japan. Reality, without the shadow of future domination by the state, is the test.

We find an even more striking agricultural advance in the Republic of China on Taiwan. One finds yields in Taiwan similar to Japan's, but attained in a much shorter period. Moreover, not only does the ethos exist on Taiwan — it was attained within a sense of peasant participation rather than solely through decree by a centralized authority. If in these two countries, within our own lifetime, why not elsewhere? The typical response, that somehow Japanese and Chinese are "different," is no answer. What is different is their agricultural ethos — and to that, all humankind will respond.

Now we turn again to the question — what can the developed world do? What should it do? Giving financial and technical aid (which is the best our imagination has allowed) to countries where the ethos does not exist is worse than subsidizing failure — it is also waste. It may even prolong the agony in that aid without the ethos is likely to be supportive not of agriculture, but of those ideological, political, bureaucratic and cultural restraints that this book will claim inhibits in the first instance. We should be quite clear and unequivocal about it all — without the ethos, no aid. This is the first great step toward development that the West can give to the planet's impoverished farmers.

There are reports of a massive fresh water aquifer beneath the Egyptian desert. Said to extend for some 500 miles and to contain water suitable for irrigation, it is believed to replenish itself by seepage from the Nile. With water levels at 2,000 to 4,000 feet, pumping for irrigation will be expensive. Nevertheless this aquifer is designated one of the planet's major resources. Combined with a favorable climate, it gives the region a potential to become one of the planet's major agricultural regions. But here we must halt.

True optimization of physical potential depends first upon optimization of human potential. The real question is: will the agricultural ethos stimulate the region's farmers, or will our inhibitors stifle them? Water has lain beneath the desert for millenia. It awaits the ethos. Meantime, where the ethos does exist — as in Japan and in the Republic of China — agricultural yields will remain the highest in the world, continuing to expand. There, physical resources may be skimpy, but human resource burgeons.

Perhaps the most striking fact about the seven points — this attempt to establish a simple agricultural ethos upon which future development might be based — is that it all has been said before. It is saddening and chastening that this is so and that it has to be said again. It must be repeated because in this modern world most farmers face a new enemy. It is no longer a rapacious monarch or a feudal overlord who denies a peasant the fruits of his work through taxes and other deprecatory actions. Today the centralized state assumes that role. The modern state has the means, through communication and the use of force, to implement those powers to a degree undreamed of by more ancient tyrants. Here lies the reason for continuing crises in agriculture. The state neither stimulates nor allows others to stimulate. The state as in ancient times seeks two synonymous things: power and money. In Third World societies these can be attained by control and exploitation of people, most of them farmers.

The task, then, is less an agricultural one than a political and, ultimately, an ethical one. Neither can advanced countries of the West avoid the politics of agriculture. To tell an impoverished country that its agricultural problem can be solved by giving it loans, new techniques or the like, without first assessing its agricultural ethos, is to beg the question. It is also a disservice. It allows the government concerned to avoid hard questioning of its own attitudes and methods. It is, in fact a pretense.

Agriculturally underdeveloped and developed nations must all face a future task. The former must look to its people as its greatest single resource and establish a simple agricultural ethos designed to optimize human capital. For developed countries a different kind of responsibility arises: it is to tell the truth.

First, we must stop telling the Third World that its abundant people are an embarrassment, a great millstone around its neck. People are assets. As Buckminster Fuller has stated in numerous speeches and articles, people are the "anti-entropy factor" in the universe.

The second truth is that the planet is not running out of resources. Particularly in agriculture, the capacity for abundance exists, awaiting its human activators.

Satellite pictures of most of the planet, soon to be improved greatly, will allow a clear defining of agricultural potential. These pictures and their interpretation should become common property, so that ordinary people can learn about themselves and, more importantly, about what they might do.

Such graphic illustrations could have a revolutionary impact, revealing the untruths that have been perpetrated upon ordinary people by their leaders and by general human apathy and ignorance. The pictures reveal that this or that country is not necessarily condemned to perpetual poverty. The pictures show that other people in similar circumstances do produce plants and animals; that soil is there and water resources await harnessing; that human capacity exists in wondrous and diverse forms in terms of solving agricultural problems.

The third truth is the essential need for freedom — freedom from ideological constraints; freedom from political abuses; freedom for attaining economic literacy and the cessation of punitive economic action; freedom from bureaucratic regulation and mismanagement; freedom for individual creativity. Agriculture in most of the world is at such a low ebb that no amount of freedom can be too much. Men often abuse freedom, but at this stage any abuse will be more than compensated for by increases in productivity.

Perhaps when entrepreneurship and good husbandry have begun to escalate productivity, regulation can return. First, however, it is essential to let people go! If one must have regulation, one should wait until the economy can afford it. It must also be recognized that there is often a psychological discomfort with liberty, especially when liberty must be compared with the rather compelling security of statism, no matter how severe. Liberty is not necessarily a total happiness factor, but it does allow things to work. To enunciate these truths will call for an appreciable re-ordering of mind for much of the West.

The West and Japan have certain other responsibilities for the future. They command the bulk of the world's scientific knowledge and enterprise. It is the West and Japan that have pushed agricultural research, development and management to its farthest frontiers. It is the West and Japan who possess trained manpower, if not entrepreneurial skills. It is the West and Japan who have created the political systems, however rickety, that have most furthered the development of human capital. It is the West and Japan who have created the wealth that makes courage and optimism a practical operational form. The rest of the world is in great part dependent upon the West and Japan. The nature of the task ahead was recently stated succinctly:

> **The crucial task of this decade, therefore, is to make the development effort appropriate and thereby more effective, so that it will reach down to the heartland of poverty, to two million villages. If the disintegration of rural life continues, there is no way out — no matter how much money is being spent. But if the rural people of the developing countries are helped to help themselves, I have no doubt that a genuine development will ensue, without vast shanty towns and misery built around every big city and without the cruel frustrations of bloody revolution. The task is formidable indeed, but the resources waiting to be mobilized are also formidable.[1]**

This book will note later some of the practical tasks that the West and Japan should and should not undertake. A primary prerequisite is that the West recapture its optimism. We must abandon the despair of the past two decades and the seeming acceptance of a fixed future, either immutable, or where material wherewithal declines. Instead we must accept the future as open and malleable by the mind and hand of man. The West must make this transition if it is to be helpful in the future to the impoverished sector of the world. The first task in any process of aid is to put oneself in order. As the Buddha noted so long ago, before one can give help to others self-help is essential. The Buddha asked, "How can one man stuck in the mud pull out another man stuck in the mud?"

There are also some other practical things relative to the future of world agriculture to which the West might aspire. These lie in four main areas. They involve tasks that only the West and Japan can accomplish. Accomplishment in these fields lies outside the capability of the Third World.

The first is that of continuing, if not enhancing, agricultural research. The food and fiber crisis as a practical issue can be met by elements and techniques we currently have at hand. Research is always necessary as long as we do not assume that some new invention will by itself solve the problem. For too long, serious analysis of the crisis has been diverted by extraordinary claims made on behalf of this and that new plant or technique. Even as I write a friend sends a newspaper clipping from New Zealand, bannered: "New Zealand Scientist May Have Discovered the Answer to World Food Crisis."[2] This is typical of the beguilement of men's minds by inventions, and the misdirecting of their gaze toward "solutions" which purportedly will ensue if these inventions ever become operational on widespread scales.

Nevertheless, the New Zealand invention looks like yet another substantial contribution to the fast developing science of plant genetics. Its progenitor, Dr. K.K. Pandey, working at the Department of Scientific and Industrial Research, has used irradiation to break up plant genes. With the breakup, it becomes possible to select only those characteristics desired. Apparently these characteristics can then be transferred to related plants. The potential that might be released by perfection of this kind of process is quite stupendous. Further, it is but one of innumerable new experiments currently under way in plant genetics.

Significant as these and many other individual discoveries may be, they mean little by themselves. Research should proceed because it presents practical building blocks for development, and not development per se. The enormous potential inherent in the planet's plant and animal life needs to be released.

The purpose here is not to make a listing of research. That would take many volumes. Rather the task is to enunciate the principle that research should be vigorously pursued in the West and Japan for potential use by all mankind. Besides the important research into plant genetics, other major plant research areas include: improvements in photosynthesis, self-nutrition through genetically induced bacteria growth, new methods of nutrition, disease resistance, yield increase through hybridization, shortened life cycles, reducing water requirements, and utilization of salty water.

Much the same is occurring with animals: genetic improvements to produce better meat, milk and wool yields; sex determination; gestation of two or more offspring; and better use of animal products, especially of what are now called "by-products."

There is soil research centered around more effective ways of carrying nutriment in soils. One might also mention the dispensing with soils altogether, such as hydroponic farming.[3] Theoretically there is virtually no limit to the amount of food that can be grown under hydroponic conditions. The process is barely mentioned in this book simply because at this moment it is a largely unneeded technique except for certain world regions such as perhaps Middle East deserts, Siberia, wartime Pacific islands or in a space colony. At this time it is usually cheaper to grow plants under "natural" field conditions than it is to build the infrastructure required for hydroponic farming. Some farmers do make a living at hydroponic farming, producing fairly high cost, but high quality seasonal fruits, such as tomatoes for select markets.

Neither will this book stress aquaculture. From time to time the media highlights such generalities as the ocean's becoming the planet's food basket. Farming oceans as a practicality is a long way off. Mankind would be better advised to concentrate for now on what he knows... his fields. Aquaculture has begun in littorals and in fresh water ponds, but it is much too soon to assess its potential. There, however, is a very definite potential, significant in that fish, molluscs and crustaceans provide protein and this is what humans demand. Further, most fish convert nutriment into protein at much more efficient rates than does any four- or two-legged animal.[4] Oceans also grow seaweed, giant kelp, the fastest growing plant on the planet and one of the most abundant. The food potential of kelp awaits man's ingenuities. Some studies envisage it as an energy source, as well.

It is regarding work (as we shall use it later in the agricultural symbiosis), that research, or should one say general analysis, has failed the worst. In terms of work by machines some spectacular advances have been made in agriculture. But in terms of work by human beings — human creativity and skillful action — research has been almost negligible.

In this area we run into man's most innate prejudices — his biases, pettiness, abhorrence of change, selfishness and lack of regard for his fellow. Here we come up against war-making man, ideological man, political man, bureaucratic man and those of man's cultural traits that bind him to nonproductive pastimes he cannot afford.

In terms of machines, however, research and development continue to produce extraordinary results; the list of products could fill volumes. From the U.S. space program come fracture-toughness tests already applied to plows and tractors, which, when translated into smaller machines such as might be used by peasant farmers, show potentialities for small-scale mechanization of dimensions as yet undreamed of.[5]

From the space program sewage treatment process becomes possible the large-scale recovery of sewage, and its use for agriculture and other purposes. Sewage represents one of mankind's most undervalued and underutilized potential agricultural resources. During a normal lifetime each human being produces about ten tons of waste. Proportionally, most animals produce more. This "waste" has a high plant nutrient value. In most countries human waste is now regarded as a vast nuisance. In the U.S. some 40% of all solids in human waste are dumped into rivers, lakes and waterways. Thanks to space technology, we now approach the threshold of making recovery practical for agriculture. Nevertheless, one wonders if cultured man, especially in the U.S., will accept what ought to be a natural phenomenon, that man should return to plants — at least in part — what he took from them.

The second area of concern for the West is climate. Here nature still reigns supreme. There have been some modifications through irrigation systems. Some plants and animals have been adapted in a very small way to handle greater climate extremes. Nevertheless, beyond devising and constructing irrigation systems, man cannot look optimistically at climate control. Not only are complex technical problems involved in true climate control — probably inordinately costly to solve; there are legal and political problems involved that today seem insurmountable. If one wishes to draw moisture from rain-laden clouds, how precisely can sought-after rainfall be allocated; and in the first place, whose clouds were they?

In Southwestern United States, irrigation farming has reached a zenith. The combination is close to contemporary agricultural perfection. The climate offers 365-day growing temperatures and ample sunshine for photosynthesis. Soils are easily cultivated. Water resources, albeit limited, for the time being offer through irrigation systems a totally controllable moisture element. As a result, agricultural productivity of what was formerly barren desert is near miraculous. The cycle goes on twelve months a year, every year, in a ceaseless pattern of growth. Alfalfa can be cut virtually continuously. Cotton or sugar beets can be followed by winter wheat and then more cotton, continuously. Great crops of lettuce can be grown when substantial portions of the nation are under snow.

In Arizona about one million acres are being farmed under irrigation. Probably another thirty million acres in the Southwest could be farmed under these advantageous conditions. We should remember that thirty million acres under a 365-day cycle is equivalent to at least double or triple that amount elsewhere. In other words, with appropriate agriculture, the Southwest could add the equivalent of yet another sixty million acres, or some 15% of America's current crop lands. But there is a problem in this rosy projection — availability of fresh water.

Taking Arizona again as an example, nearly half of the state's water is drawn from great aquifers underground. Such exist virtually all over the planet, and offer distinct advantages. Although much of this underground water is lost by percolation, depending upon the permeability of surrounding soil structures, the rate of loss is very much slower than a loss through evaporation (as with above-ground water). In other words, underground is an excellent place to store and preserve agricultural water. Certainly there is a cost involved in extracting this water. Wells have to bored, usually several hundred feet deep. Large pumps have to be installed. A distribution system from well head to field has to be created. A supply of energy to activate the pumping (and sometimes the distribution system) needs to be at hand. If one invokes "flood irrigation" (distribution by canals and ditches with simple plastic pipe siphoning into each row of crops), these fields must perforce be first levelled by graders and other machines. If one distributes water by various spray systems, large circular arms known as pivot-point sprinklers being popular, these too have to be manufactured, installed, and paid for. But the cost of irrigating is balanced by rewards of increased productivity.

Where water is drawn from aquifers, often these aquifers are either not recharging at all, or if so, at a lower rate than the extraction. It is possible that some water being pumped in the Southwest, for example, goes back to the last Ice Age. Some of these vast aquifers will last for hundreds of years, but ultimately their resources will be expended. Then if nothing further is done, desert will return. Great cotton lands, alfalfa stands, citrus trees, pecans, and near endless varieties of vegetables and fruits will disappear. In their place there will be a jack rabbit or two per acre, a few lizards and reptiles, fewer small animals. Perhaps a hardy rancher will run a mother cow per one hundred and fifty acres. Mankind will have lost a resource.

There is, however, no general shortage of fresh water on the North American continent. Neither is there on the planet at large. The enormous above-ground fresh water resources of American rivers and lakes represent only about 5% of this nation's fresh water resources. The remaining 95%, underground, is partially recharged by runoff from rainfall. This underground water is impervious to drought. One U.S. Geological survey report estimates that, based upon its present rates of usage, the arid State of Arizona could withstand more than a 100-year drought. There are, however, as with other natural resources, flaws in this pattern of resource abundance.

Again taking Arizona as one end of a spectrum of water availability, it is estimated that some 95% of the state's small annual precipitation of rain and snow either evaporates or percolates away. Only 5% is captured in natural aquifers and surface dams. To redress this imbalance by even 5% would therefore double the state's annual water recapture rate. At the other end of the spectrum, some other states have too much water, especially with thawing of winter snows or severe rainy seasons. This water drains off and eventually reaches the ocean. The problem of effective water supply and usage is, therefore, one of distribution — actually re-distribution.

There are strong arguments against U.S. water redistribution schemes, but not for the rest of the planet. The economics of present U.S. agriculture are such that massive food and fiber surpluses can be generated without the high capital expenditures needed for water relocation programs. The situation will not, and ought not, remain this way even in the U.S. For much of the rest of the planet, fresh water distributed to crop lands is a must. The luxury that the U.S. enjoys, being able to disregard utilization of a food-producing resource, may disappear as local regions turn arid because local water is expended, and especially because rising affluence in the rest of the world may place real economic pressure on America's capacity to produce food.

Today certain tiny segments of the U.S. population can halt dam construction or river diversion because of concern for the future well-being of a tiny minnow or because a bird variety may be forced to change its nesting habitat. It is as if natural habitat for life forms on this planet had never changed and, indeed, can no longer be permitted to change... as if evolution must be brought to a halt and nature solidified in its present form. Obviously this is nonsense. Nature will go on changing whether man is involved or not. All life forms are subject to death and replacement. Man stands a good chance, however, of adapting to change, because of his intelligence.

We can be thankful that this movement to halt natural evolution has been with us for only a very short time. The dinosaur, as but one example, was apparently an extremely specialized beast with an equally specialized but elementary digestive system. Apparently it consumed large quantities of quite limited types of plant life. One speculation is that when these forms of plant life disappeared, so did the dinosaur. One can be thankful that our environmentalist friends did not exist in those times. Perhaps if they had we would now be involved, under quite impossible difficulties, in supporting dinosaurs. The same argument can be applied to tens of thousands of plants and other living forms.

In the face of evolving nature, attempts to stay evolution become quite ridiculous. It will be argued however that "natural" evolution is acceptable. It is the destructiveness of man that has to be halted. Man *has* been destructive. Primitive men, such as the "slash and burn" agriculturists, still are. But let us assume that many men, including farmers, have intelligence. It makes no economic sense for man to preserve dinosaurs. We have much to learn in this regard. The environmentalists have rendered service to humanity in raising the level of consciousness in this regard, but we need to exercise common sense.

In nature there is always a trade-off. Nothing is ever pure gain. To attain something, something else must be given up. A dam may be worth the extinction of small fish species where the benefit of the dam is man-centered rather than fish-centered.[6] One can become sad at possible extinction of Alaskan caribou, following near-extinction of the bison. But the bison has been supplanted by approximately 135 million beef cattle and 12 million dairy cows. These herds do not face extinction (thanks to the virtues of private as opposed to public ownership).

To return to climate and man's attempts to come to terms with it, the greater bulk of the planet's people cannot indulge in the luxury of saving small minnows. They must catch minnows for food. For them, control of water and redistribution of water is yet another key, albeit a physical one, to their agricultural future. Only by the hand of man can natural maldistribution of water on the planet be reversed. And it poses problems. Changes in ecological balances will occur. These must be judged primarily on a cost-effectiveness basis. There will always be a trade-off. The key is to understand the consequences. We simply cannot afford to shout, "Stay, proceed no further; for the Southern eagle will now have to develop new nesting patterns and this is unacceptable." Perhaps the Southern eagles's plight is part of the cost. The trade-off may be to fill the dam with catfish, trout, bass, bluegill, crappie and carp. One will also be amazed at the bird life that will follow.

Beyond this is the alternate use of water in terms of human welfare at home and abroad. Vast areas of arid, nonproductive lands can be made to double production comparable to lands fed by natural rainfall. It also means that because of unpredictable weather patterns, land management can be comparable to industrial production management, with commensurate savings in energy — human and mechanical. Irrigation projects are costly; but because of their long life under normal economic conditions, the return on investment ought to be excellent. While inflation reigns, however, potential return in investment based upon long-term gain may disappear. Inflation is symptomatic of a far deeper malaise in modern society. Without irrigation systems, agriculture will either not advance or will decline and living standards will further decline, hence the task is not to reject building irrigation systems because of long-term loss due to inflation, but to get rid of inflation.

Another cliche is "people should be moved to water and not water to people." The objection is primarily directed against the U.S. "sun belt," where arid regions become inhabited and thus increase water demand. And there are other objections. From an agricultural point of view water ought to be brought to the most favorable soil and climate regions. Montana obtains substantial water resources each year but its temperature limits the growing season to about four months. In contrast, the irrigated desert can blossom year round. Climate is quite intractable. It cannot be shifted around. Water can. It is natural for people to follow.

A fundamental right is to choose where one will live. Yet in some circles in the U.S., this concept is under attack. Similarly, those who enjoy the "sun belt" sometimes try to keep others out. People are told they are a liability. There are those who presume they know what is best for others, and would legislate the right of people to move. As it happens, however, people regulate themselves rather well. Beyond this lies always a fundamental question. Do people exist to serve the environment, or should not an environment exist to serve people?

Control and movement of water is yet another task that must be directed or assisted by Western nations. The Third World has only the most primitive skills in this regard. The West has experience and engineering skill to accomplish great engineering feats. The task of further extending irrigation to arid regions of the planet is primarily a Western task.

There is, too, something philosophical about irrigated farming. When, for example, one pumps from great aquifers water thousands of years old, and ultimately expends such water, one can wonder at the rightness of it. Certainly, over future thousands or millions of years, the water would probably disappear. Already vast stores of water from the last Ice Age have percolated into nothingness. Today we use but a tiny fraction, a residue. But where there is no recharging, in a hundred years, two hundred years, it will be all gone. Nevertheless, by use of this water something profound will have happened. Non-productive land can be turned into some of the world's most productive land. The desert will rejoice and so should man. He has demonstrated how far mankind can advance from a hunting and gathering status. It is an affirmation of man's creativity and innovation, his skill and his work. It also offers visions of a future. A demonstration has taken place.

The task, therefore, is not to wring one's hands about water that will soon be no more. The task is to undertake those great works that will create, conserve, retain and redistribute more fresh water.

A map compiled some years ago reticulates water resources of the North American continent into a vast, interconnected grid. Rivers are moved over mountains, rivers are divided, rivers are co-joined and become great new rivers, new great lakes are formed. Basically, it is a vision of collecting water and moving it southwards, from colder to warmer regions of the continent. Sometimes we search for national goals. Here is one. It calls for work and investment on a scale as yet not attempted. It calls for human ingenuity and innovation. In the U.S. it calls for political cooperation between states. It calls for new kinds of cooperation between all countries on the North American continent. It allows for creation of new aesthetic landscapes.

What could be done on the North American continent, exemplified by this vast new demonstration, could be done elsewhere on the planet. Mankind might reconsider his options. Instead of worrying about hunger, we might instead turn our planet into a garden.

Again, before we leave water, let us return to our tiny minnow doomed to extinction if the dam is built. Intense concern over the fate of some tiny fish presages more than mechanics of the fish's future. The late Paul Tillich talked of symbols.[7]

> **Symbols, although they are not the same as that which they symbolize, participate in its meaning and power.**

Thus, may not the symbolic meaning of this seemingly impractical concern with minnows symbolize among us a giving way to nature itself? Tillich goes on:

> **Out of what womb are symbols born? Out of the womb which is called today the "group consciousness" or "collective unconscious," or whatever you want to call it — out of a group which acknowledges, in this thing, this word, this flag, or whatever it may be, its own being.**

If our own being is to be the object in the equation with Nature, and our own being is not to be the subject in the equation, then the future of agriculture, and thus humanity, is dark indeed. When man ceases to be the subject in the universe, Nature will roll inexorably over him, wreaking immeasurable personal tragedy as it does so. The small minnow is a symbol of a society that stands to lose its sense of proportion if not its attitude to right and wrong. It is more important, perhaps, as a symbol of our future than even environmentalists think.

The third main area of concern to the West is co-joined with water, and that is energy. Water cannot be controlled or redistributed without energy, nor can farming advance. The planet needs a plentiful, if not unlimited, supply of cheap energy. This it has had to date with wood and, in modern times, with fossil fuels. As irrigation gives dramatic demonstration, fossil fuels helped show the agricultural potentials that lie before us. Productive agriculture consumes large quantities of energy. The highest user of energy per agricultural unit is the U.S. This is to be expected because the U.S. also has the world's highest per-man-hour agricultural productivity. In other areas of high agricultural productivity, such as in Taiwan and Japan, the consumption of energy per unit production is less than in the U.S. because they rely more upon human power, but consumption of energy in such sophisticated peasant agriculture is still high. There will be no agricultural advances without new sophistications among the three billion plus who comprise the world's peasantry. Thus they too will need cheap energy.

In the U.S. it takes about ten calories of energy produced from oil or coal to produce one calorie of food. Measured in this regard, agriculture seems quite inefficient. A ten-to-one ratio represents inefficiency in almost anything. In the midwest a farmer uses about eighty gallons of petroleum to produce an acre of corn, yielding 150 bushels per acre. In 1945, only about ten gallons of fuel was required but the yield was much lower. Of the 80 gallons currently used, about 20 to 25 gallons is used by farm machinery. The remainder primarily goes to produce fertilizers, usually nitrogen, phosphorous and potassium. Beyond the farm, energy is consumed by processing, at perhaps twice the rate used for growing. Transportation consumes energy at about one-tenth of the rate for growing, while wholesale and retail services use about the same amount as does growing.

Overall, one American consumer requires close to 30 million B.T.U.'s (or calories) per year in terms of food and fiber and related energy input for his sustenance and shelter. In more equitable terms, every human being on this planet deserves approximately the same expenditures if he is to enjoy what life should give him in return for his skills and labors. For some reason, such data often fills people with despair, and provokes them to declare that there are too many of us or that we should go back to some primitive form of life. There is, however, no tolerable life form requiring small amounts of energy. One learns this on one's first day in a low-energy society. The only solution to a demand for energy is to create energy and insure that it is created as cheaply as possible.

We live in a universe the essence of which is energy, yet to date, mankind has taken but a few halting and primitive steps in harnessing the energy of the universe. Electricity, that potent force that no one even yet quite understands, is by and large generated today through the combustion of fossil fuels. It is fossil fuels that provide the greater bulk of all other forms of energy demand. These fuels are finite and may soon be expended. As shall be noted later, neither is it tremendously significant in history whether this occurs in the next fifty years or one hundred years or as is more likely in 1000 years. It is quite clear that when one matches the need of a planet where each human requires 30 million B.T.U.'s for adequate living standards, present energy configurations will not be adequate. Again we need to let creativity flow. Beyond fossil fuels, we know of many sources of energy that can provide for human needs. If we were sensible we would use fossil fuels to bridge the gap to allow us to get there. We would also take what has been achieved by energy generated by fossil fuel as grand demonstrations of what can be achieved in the service of man by exploiting energy. We can draw energy from the sun, and perhaps through the proper use of space satellites. We know a great deal about fission. As a priority we should move toward fusion energy as a nearly inexhaustible source. Allied to fusion energy is the possibility of deriving energy from deuterium in sea water...another near-inexhaustible source.

On the other end of the scale we should look at other simple energy forms. It is estimated that in northeastern U.S. alone about 3,000 small dams have been abandoned. These could all generate electricity for local needs. There are often, but by no means always, surprising economies to be had in small projects. There are wind power and tide power, with a potential to serve local enterprises. From a farmer's point of view, one exciting — if yet economically unproven — energy source is bioconversion of plants themselves. Each plant on this planet is basically a tiny solar energy converter. Part of its energy can be recaptured, mostly in the form of ethyl alcohol. Sugar cane, potatoes, sorghums, corn, wood and the wastes of these are some of the basic raw materials. The point at issue is that of bending our imaginative minds to the goal of creating cheap and abundant energy, a more exciting prospect than deploring a fate conjured up from our ignorances.[8]

The future task — and it is enormous — of harnessing new forms of energy can be done only by the industrialized West and Japan. The record of U.S. political leadership has, however, been anything but reassuring. Measured in terms of its pusillanimity, lack of vision, rhetoric, encouragement of energy waste and discouragement of energy creation, the record is quite alarming. The potential of planetary agriculture will perhaps be determined by this capacity, or incapacity, on the part of the West and Japan, and particularly the United States.

In synthesis, abundant energy can be had for all. Without it, real agricultural progress will be difficult, if not impossible, to attain. Water projects will founder. Mechanical energy in the field and plant fertilizers will become expensive and scarce. Food processing, transportation, wholesaling and retailing will languish. The cost of ordinary food staples will increase; perhaps some will disappear.

Some romantics believe that somehow, in some new Arcady, life will be pleasant. They talk of organic fertilizers, nature's bounty of rainfall, manual labor as a substitute for mechanical labor and the like. All these features can pertain — and should. The Arcady, however, will be less than idyllic, and by an awesome margin. Anyone who has lived in a peasant village in a traditional agricultural environment will rapidly perceive the true dimensions of Arcady without energy. It is really quite miserable, dull and enervating. Even the most romantically-minded Arcadian would not like it.

A new sophisticated breed of small farmers (which has to be the mode of development in the Third World) must of necessity be high consumers of energy, using probably not too much less than today's American farmer. Their fertilizer requirements will be exactly the same. It makes little difference whether a farm is two acres or 2,000 acres. The energy per acre, in terms of machines and fertilizer, is the yardstick. Increased human-power can substitute for a portion of the mechanical energy that is expended on an American farm. It cannot substitute for much else if high yields are to be attained.

The energy requirements for food processing, transportation, wholesaling and retailing will be about constant all over the planet if every human is to have a variegated, nutritious diet. Precisely the same argument can be made relative to energy expenditures in production of fiber for clothing and shelter. Thus cheap and abundant energy is a future necessity. We have the opportunity to obtain abundant and cheap energy. It is the particular task of the West and Japan to accomplish this.

The final area of primary Western responsibility is that of more esoteric and imaginative operations. Pioneer 10 is on its way to probe deep in space. In 1973 this small unmanned space satellite was aimed at the planet Jupiter. In December, 1974, Pioneer 11 was renamed Pioneer Saturn and flew within 26,000 miles of Jupiter and proceeded on to Saturn. It passed through the rings of Saturn in 1979. Pioneer 10, on the other hand, will head on a course that will take it out of our solar system. In doing so it will pass the orbit of Pluto in 1987. Some 4,000 years later Pioneer 10 will coast the immense distance to the nearest star in the direction of the constellation of Taurus the Bull.

Pioneer 10 carries within its 750-lb. weight a tiny gold plaque, six inches by nine inches. The plaque is designed to communicate with intelligent life. Its simple message tells of man and from whence he came. On the right hand side of the plaque are figures of unclothed male and female humans, the figures in proportion to the size of the space craft. There is also a diagrammatic representation of the hyperfine transition of neutral atomic hydrogen, considered to be a universal "yardstick." There is a representation of the sun relative to 14 pulsars and the center of the galaxy. The planets of our own solar system are also represented together with binary relative distances. Forty centuries from now Pioneer 10 will be in regions where there could be planets not too dissimilar from Earth. Beyond that, over distances beyond the scope of human imagination, lie vaster galaxies, hundreds of millions in number and with hundreds of millions of planetary members in each. Pioneer 10 is symbolic of man's first tiny probe of outer space. In 1977, Voyager, a much more sophisticated version of the early inter-galactic probes, went aloft.

If one accepts an elementary notion of probability, there is little doubt that in this immensity are planets equally favorable to man as is Earth. Further, there seems to be evidence that within this awesome Universe there is a continuing process of death and rebirth of galaxies, a process of continuous creation and recreation. In any time scale remotely conceivable by man, presumably the Universe never dies. The question thus becomes: can man recreate himself throughout the Universe? Under present conditions he cannot because he lacks the means to maintain life under deep-space-probe conditions. But he has made the first step. The future will offer optimism as he takes other steps, steps of unimaginable dimension. As in the line by e.e. cummings, "There's a hell of a good universe next door; let's go."

It is possible, too, that in the very long term, farming soil under natural conditions might disappear altogether. Because of the basic configuration of the planet's biosphere, and especially because of the inchoate climate, farming as it is practiced today is an inefficient enterprise even under optimum conditions. One productive unit in agriculture requires about ten units of energy for its realization. Extensions of the hydroponic farming technique under enclosed and designed optimum conditions could allow evolution of a radically more efficient agriculture. Protein might also come from chemical constructs rather than animals. Space colonies currently being contemplated illustrate this potential.[9] To the joy of environmentalists, the landscape might then return to a pristine state. Those messy human activities associated with tractors, cattle and corn could be confined to antiseptic factories.

We need to lift up and to reorder our imaginative minds. We can do this. One of the extraordinary things about being human is our ability to imagine. We all have it. I recall in the late 1950's coming in contact with a most humble group of people. It was a tiny band of less than twenty souls. Their role in life was to drive water buffalo from the Korat Plateau in northeastern Thailand to the port of Bangkok. Here the animals were shipped to other Southeast Asian countries and to Hong Kong. Thai water buffalo are greatly prized as work animals.

I came to know the headman quite well. What a remarkable man he was! He could neither read nor write but he was a natural mystic, a religious man in an animistic sense. He would tell me of the spirits that inhabited forests, water, stones, houses and all animals. He possessed an ethos for the buffalo which he expounded upon at length. He was interested in my description of Western cattle enterprises, but he felt that I was too practical and materialistic. I did not know the mind and soul of my animals, he said.

I parted from the group and returned to Bangkok. Months later when driving to the airport I noticed a long rank of buffalo, tied by their noses to a picket line. I thought I recognized the animals. I alighted and soon found my friends. They had built tiny thatched houses of straw and were awaiting transshipment of their charges. The headman squatted outside his little house, smoking. We began to talk. At that moment a Comet jet transport took off from the airport. It skimmed very low over us. The buffalo shrank to their bellies. One near me even uttered that short rasping bark so typical but so rarely emanating from buffalo. They do it when terrified.

When the jet had cleared I commented to the headman that the machine would be in Paris or London in not too many hours and was carrying perhaps 100 persons. Did he not, I asked the headman, think that a rather remarkable feat? He removed his cheroot and began to talk. For me he opened a new vista. It was a marvelous exercise in virgin imagination.

He told me of a world of magical flying machines, transcending time and space. The machines had animal, bird and reptilian shapes. Humans were transported to new realms and so took on new dimensions. The new worlds opened up by these fantastic machines gave new lives to humans. It was a melange of marvelous imaginings and Blake-like visions. It was man recreating himself in the Universe. It came from a humble illiterate cattle driver. Yet it was more descriptive, if not a more impressive forming of mental images and concepts, than I had ever heard from any educated man. This is what we need to recapture.

For the last thirty years at least we have tended to try to solve our various difficulties by political and administrative means. As a result we have come near to creating a world of apathetic peoples half enslaved to unresponsive governments. We have lost sight of the worth of the individual as innovator, creator, manager and worker. We have assumed, wrongly, that he was only an ideological and political animal. Unless the individual is once more put at the apex and institutions made subordinate to him, it is unlikely that the vigor and enterprise needed to regenerate and enhance the agricultural symbiosis will be effected. The price that we shall pay for this default will, in agriculture, be continuing chronic impoverishment for billions. But we can avoid this if we accept and proliferate the agricultural ethos espoused in this chapter.

To shake free from that which inhibits agriculture thus, in the end, becomes the primary human task for the future. We must examine agriculture's inhibitors in detail. Above all we must reorder our imaginative minds. The potential ahead of us in this instance becomes magnificent and exhilarating.

FOOTNOTES

1 E.F. Schumacher, *Small is Beautiful* (New York, Harper & Row, 1975), p. 204.

2 *The Evening Post,* Wellington, New Zealand, Dec. 30, 1976.

3 To illustrate, with one example, the kind of farming research under way (and in this instance well developed) the nutrient film technique might be described. Water in which nutrients are dissolved is reticulated from an accumulator along a channel and back to the accumulator or reservoir. On the way this nutrient-laden water envelops plant roots and nourishes them in accordance with the plant's biological need. The plants are supported but their roots dangle in the water. A chemical analyser constantly monitors the water and adds the appropriate amount of nutrient as needed. A refinement, especially in desert areas, would be to use solar heat under a dome to condense sea water to obtain the fresh variety.

4 One of the most interesting aspects of aquaculture concerns the current "save the whales" controversy. Whales are an excellent protein source. Their yields of meat per carcass as a percentage far exceeds land-bound mammals. If whales could be farmed, it is suggested there would be no more need to agitate to "save the whales" than there would be regarding the "saving" of Herefords, Angus or other beef cattle, of which the U.S. alone has about 135 million.

5 For a general survey of space technology and its general application, see National Aeronautical and Space Administration, "Spinoff 1976, Technology Utilization Program Report," Washington, D.C., April 1976.

6 The real challenge for the future is not to build dams with their high evaporation rates, but to learn how to refill aquifers and other holding areas underground where water loss is generally much less than in above-ground storage, and to pump directly from lakes and rivers.

7 Paul Tillich, *A Theology of Culture* (New York, Oxford University Press, 1959), pp. 54 and 58.

8 A compendium of references to work currently under way in various forms of energy conversions is given in the general bibliography.

9 T.A. Heppenheimer, *Colonies in Space* (Harrisburg, Pa., Stackpole Books, 1977).

THE AGRICULTURAL SYMBIOSIS 2

This planet is a potential agricultural cornucopia capable of giving food and fiber abundance to each of its four billion inhabitants. The fact that it does not is a planetary crime.

In one study it was estimated that if full exploitation of currently known arable lands was made, the planet could feed, shelter and clothe 35 billion people using Western European standards of diet as the yardstick. It was further calculated that if known arable lands were made to bear the same yields as currently occur in Japan, and using Japanese standards of diet as the measure, then the planet could feed 100 billion people.[1]

In 1976, *Scientific American* devoted an entire issue to planetary food and fiber. This study indicated that food and fiber production could be increased twelvefold. This would be sufficient to feed adequately, shelter, and clothe some 40 billion people.[2]

The estimates were based upon known arable land potentials and currently employed agricultural techniques, not including aquaculture or hydroponic farming. Land availability, however, is constantly changing. Today's desert, when placed under an appropriate irrigation system, could become highly productive. Such areas may become the most productive farmlands in a region. In other instances, salt-laden lands may become productive through genetic plant changes, rendering them less vulnerable to salty soil.

On the other hand, there are factors which reduce the area of land available for farming. All over the world productive agricultural lands are eaten up by urban development. In other instances, lands are losing their productivity either because of poor farming practices or because of inadequate fresh water. Thus, the task of estimating availability of arable lands involves a dynamic situation where it is best to avoid certitudes.

Nevertheless, we should note two important conditions. The first: as we have improved agricultural techniques, so have we broadened the spectrum of lands which can be farmed. For example, the advent of tractors made it practicable to farm lands hitherto unfarmable. Indeed, most every advance in agricultural technology, agricultural management, fertilizers and their network of application, and in plant and animal genetics, has opened up the potential for farming new lands.

The second condition: we know there are vast areas of the planet which could be farmed under current conditions, but are not being farmed. The following simple table illustrates this:

TABLE 1[3]

Region	Percentage of total land area which is potentially arable (approx.)
North America	50
Europe	12-15
South America	88
Africa	85
Asia	18
U.S.S.R.	35
Oceania	90

A second table illustrates another interesting fact: the availability of tilled land per person. This statistic is especially important when we later compare yields per unit area in various countries and the number of people supported by a given area.

TABLE 2[4]

Country	No. of people per acre of tilled land
Japan	7.4
Netherlands	4.4
Egypt	3.9
U.K.	2.9
China	2.5
Peru	2.4
Indonesia	2.0
Norway	1.7
Brazil	1.4
Italy	1.2
India	1.1
World	0.8
Mexico	0.7
Nigeria	0.6
U.S. and U.S.S.R.	0.38
Argentina	0.28
Australia and Canada	0.17

But the availability of land is the least significant element in estimating new agricultural potentials. The greatest potential lies in increasing yields per unit area.

Thus, we can assume that the planet is currently producing below its optimum, but has as yet unexploited capability to yield substantially increased food and fiber to its present population — and more. Optimum productivity is less a matter of acreage yields than of more complex human factors.

Agriculture is man's oldest occupation. Here is where we must begin — with man, the way he behaves, and with his unchanging need for food, clothing and shelter. To accept evolution is to accept the fact that man is still evolving. Perhaps in this continuing evolution man's needs for food and fiber will change. But whatever evolution brings, substantial changes in man's dietary needs are probably a long way off.

An individual needs about 2,500 calories or a pound of food per day. The ideal chemical balance for good nutrition is complex. A fundamental factor is the percentage of amino acids ingested as protein. In turn, protein so ingested is retained as human body cells, which again are protein. Requirements vary according to a person's personal and environmental hygiene, internal parasites, daily temperature, and type of clothing and shelter. Over all, however, the requirement is fairly standard. Let us look at a simple but well known table that specifies an ideal intake.

TABLE 3

A Recommended Daily Allowance

Age	Sex[5]	Requirement Per Day	
21 years	Male	Protein	- 37.5 grams
		Vitamin A	- 780 micrograms
		Vitamin D	- 2.75 micrograms
		Thiamine	- 1.2 milligrams
		Riboflavin	- 1.8 milligrams
		Niacin	- 18.8 milligrams
		Folic Acid	- 180 micrograms
		Vitamin B12	- 2 micrograms
		Vitamin C	- 30 milligrams
		Calcium	- .42 grams
		Iron	- 20.5 milligrams

Each day the biosphere, that thin layer of soil and air that covers the planet in about the same proportion that skin covers an apple, must produce this dietary requirement for each human being. This means production of grains and cereals, vegetables, legumes or beans, nuts and other seeds, dairy products, and flesh from animals, fish and birds.

The biosphere must produce more than two million tons of food, or about one trillion calories, daily. The alternative to this requirement is malnutrition, or for some, death by hunger. Therefore, we need to know what the planet can produce and what man must do in conjunction with nature to so produce. This is the first essential step not merely in nurturing our bodies, but as the basis for building a civilization.

Man was not always a tiller. Up until late Neolithic times man was a hunter and gatherer. There was, however, very little symbiosis between man and nature. For whereas nature contributed everything to man's support, in turn man provided nothing in support of nature. No matter how romantically atavistic neo-Rousseauians may extol man in his "natural" state, man as a hunter and gatherer was both a parasitic exploiter and a potential victim. As a parasite he killed or picked regardless of an object's seasonal cycle, maturity or immaturity, or progenerative status, and without regard to conservation or economic circumstance. It is estimated that each person in a hunting, gathering society required anything from four to seven square miles of territory for survival. Thus the continental U.S. had the capacity to support from 400,000 to 750,000 persons. Man paid a price for his parasitism. He was subject not only to nature's whims (which are rarely gracious) but was also doomed to whatever status nature's bounty conferred. Man in this state had little to do with his own destiny, but this changed.

With the advent of agriculture, man brought about the possibility of symbiosis with nature. Man and nature lived together and contributed to each other's support. Man's work, his husbandry, created the relationship among plants, animals, soil, climate, and man. It is an intricate relationship, as we shall observe. It is dynamic and changing even though many elements remain in old forms. Nevertheless, man's ability to relate to these elements and then to relate each to the other in productive ways must represent one of humanity's greater triumphs. Man remains dependent upon some plants and animals. As agriculture evolved, man helped plants and animals to become more productive in resisting disease, other parasites, and climate. In turn, the increased productivity of plants and animals helped man become a stronger being. This process, far from being ended, is now beginning to reach a most exciting and fruitful stage. This results from genetic strategies, better plant and animal nutrition, disease resistance and control, inter-relations between agriculture and other technologies, and through management techniques. Tardy and hesitant as it may have been, the revolution which began with the animal-drawn plow of late Neolithic times is now coming to its maturity. The critical factor in the process has been man.

Husbandry is a fine Old English word for administration and management of a household and its supporting unit of land. It includes tillage and cultivation of the soil, breeding and rearing of livestock, and preservation and extension of these arts over time and space. It is also a state of mind. In 1628, Preston, a little known English writer noted, "How goodly a sight it is when a man looks into the husbandrie, to see the vine full of clusters, to see the furrowes full of corne." His contemporary, Blithe, was a little more pointed. "Consider the vast advantage there will be by Husbandring a little well . . . One acre Manured, Plowed and Husbandred in season, may and doth usually beare as much Corne as two or three ill Husbandred." The spirit of husbandry is still the keynote of effective agriculture. It is concerned with totality of the earth's biosphere and plants and animals that inhabit it. It is the beginning point, the essence of the human dynamic of agriculture — of symbiosis.

Husbandry is no less essential in modern agriculture than it was in the time of Preston and Blithe. Two men can look at a field of cattle. One man will spot the animal with a pending illness. The other will not. Two men can look at a field of grain, and the weather. One will change his plan and harvest early. The other will have thunderstorms flatten his crop. Two men will study the market. One will change his crop pattern (or hold to it) and succeed. Two men will seek financing and one will integrate these funds with an overall tillage concept and will burgeon. The other will regard financing as a loan to be endured, a burden repressing his creativity. One man is a husbandman and produces bounty. The other is not and produces less.

The husbandman has a fellowship with plants, animals, soil, climate, and work. This gives rise to their greater development and yield. It is a fellowship as natural, and with as comparable a warmth, as exists between human and human. But it is obviously different. This difference leads to a very important characteristic common to all good husbandmen. In the analogy of the Cave, portrayed by Plato in his *Republic,* one feature was that of seeing things as they really are, as wholes, not as poorly etched images. Plato's Greek word summarizing this state could best be translated as obtaining "The correctness of the glance." When a good husbandman looks about him — in field, marketplace, or bank president's office — this is what he has: a "correctness of the glance" — seeing things in truth and therefore as they might be. Husbandry at its height is an art.

Man as an organism is intrinsically related to plants. He is dependent upon them for his survival. Plants are the basis of the earth's food pyramid, food not only for animals and humans but for other plants as well. Basically plants perform the most important single chemical process in the planet — and perhaps in the universe. The process of photosynthesis, whereby light energy from the sun is transformed into chemical energy, is a miraculous event, on an immense scale. The planet's green plants alone produce up to 150 billion tons of different kinds of sugars per annum. In turn these sugars are the main energy source for all plant and animal cells.

This dwarfs into insignificance any man-made activity. The planet produces less than 400 million tons of steel per annum; human production of any other manufactured product (except for foodstuffs) is in declining proportion. As we shall see later, this remarkable generative activity known as plant life is dependent upon light, gravity, temperature, water, chemicals and a relationship with other plants and with animals. Plants are living creatures as we are, but different in their characteristics.

Plants, rooted in one spot must adapt themselves to their specific environment. They are vulnerable to any change in that environment — lack of moisture, changes in temperature, other climatic disturbances such as high winds or heavy rains and floods, or to marauding insects and animals. This is where man comes into the plant's life cycle and indeed, into survival of plants as a genre.

As has been noted, plants in their natural state provide a poor base for support of human life. Even if roving animals were added as a potential food source, these, together in their natural state, would still provide a limited supply of food for man. It is probable that man did live as a hunter and gatherer for many millennia. We know from the few humans who still exist under those circumstances that Hobbes was more right than Rousseau. Man "in a state of nature" may not live a life that is totally "solitary, poor, nasty, brutish, and short," but he does (or did) live rather close to this state. As a hunter and gatherer his capacities were highly restricted.

One small example, drawn from today's world, well illustrates the plight of the gatherer. It has been estimated that in the Maharashtra region of India, peasant families may spend up to 200 days per year gathering firewood and dung for fuel. In terms of cost effectiveness this is an extremely poor return of energy for energy expended. A similar pattern exists everywhere hunting and gathering still exist. It was no better in the past. One person gaining his sustenance by manning a four-to-seven-square-mile tract was in a somewhat impecunious, if not fragile, situation.

On the other hand, by working with nature and by changing nature, man has now accumulated under his husbandry more than one thousand plants which are used as food and fiber in various forms. Besides plants consumed by man, there are food plants consumed by animals, whose flesh, milk and other products are used by man. Animals may also work for man. As well as general food plants such as grains, vegetables, legumes, nuts and seeds, which man uses for sustenance, we may also include some exotica. These are plants which produce drugs, oils, cork, rubber, resins, beverages, and the like. Then there are fiber plants which we can define broadly as producing clothing and shelter. Lumber, bark, fronds, flaxes, sisal, hemp, and particularly cotton are representative samplings.

Man's own mobility has also introduced another important factor — his migration. Through exploration and subsequent development of communications between diverse parts of the globe, man has spread plants on a scale and at a rate hitherto unknown. Barley, oats, wheat, peas, red lentils, and rice were taken to the Americas from Europe and Asia, and further afield than that. Corn, white potatoes, sweet potatoes, squash and beans from the Americas were sent to Europe and elsewhere. These are but the main elements in the migration. The migration goes on today and in terms of more exotic crops, is probably accentuating.

Genetic improvement of plants, almost totally achieved by man, has produced two distinct advances: variation of species has been spectacular, but the increase in yields has been even more so. Variations of species are too numerous to delineate in full. As one example, a wild cabbage gave rise to a mustard variant. This mustard plant has now in turn given rise to common green and red cabbages, cauliflower, broccoli, Brussels sprouts, rutabagas, kohlrabi, and the white turnip.

The increases in yield have made the greatest advances. The following table, covering common food plants in the U.S., illustrates this point.

TABLE 4

YIELD INCREASES IN COMMON PLANTS

YEAR	WHEAT Bushels/acre	CORN Bushels/acre	COTTON Pounds/acre	RICE Pounds/acre
1866	11.0	24.3	121.5	—
1901	15.0	18.2	168.2	—
1942	19.5	35.1	272.4	1,996
1975	30.6	86.2	453.0	4,555

It should be noted however that these enormous increases did not arise solely through genetics. An important factor was improvement in plant nutrition, primarily by the use of fertilizer. The two work together and can be cumulative. For example, from these and other causes the output per man hour on U.S. farms increased ten-fold from 1870 to the present.

Another equally significant aspect of genetic development has been creation of disease resistant plants and adaptation of plants to differing soils, climates and farming methods. Nor is this process of genetic selection at an end. On the contrary, it is a relatively new science with a great future. Further, as we shall see shortly, the genetic advances of plants have been allied with genetic advances in animals, again producing cumulative results.

But there is a trade-off, as there always is when man and nature relate. Now that we till, sow and nurture rather than gather and hunt, some 40% of our man-enhanced cultivated plants, involving more than 80% of our croplands, are annuals and must be resown each year on newly tilled land. The layer of fertile soil, or topsoil, on the planet's skin is extremely thin. When this soil is exposed to tillage every year, it is subject to erosion by wind or rain. More seriously, and especially in heavy rainfall areas like the tropics, exposed soil is subject to leaching. Here nutrients are washed out of the soil, reducing its fertility, which can lead to disaster.

One theory is that the fascinating civilization of Angkor, in what is now Cambodia and which flourished from about the 5th century A.D. to the 15th century A.D., in part suffered its demise this way. Angkor's cultivated lands, which were heavily cropped — more heavily in fact than they are now — were of laterite composition. Such soil, if not fairly continuously protected by a plant covering, can turn into a light sandstone, a process greatly accentuated by hot tropical sun. This happened at Angkor when, presumably through war and other man-made neglect, these laterite soils were left untended for periods. A similar deterioration of laterite soils has occurred in Nigeria and in Brazil.

To protect soil — any soil — against erosion is a necessary consequence of annual plantings. The key is husbandry. It is the eye of the farmer correctly adjusting his crop cycles, mainly to weather phenomena, which is the focal point. But no matter how skillful, the farmer will not be successful every time. Again husbandry comes into play. The ravages of one year must be repaired in ensuing years.

Further potential loss of genetic substance arises from man's increasing expansion of cultivated lands. When wild plants in their natural state are destroyed, we lose a "plant bank" of near infinite variability from which we might draw species for further plant improvement or to produce new crop plants. We cannot foresee just how important new plants may become in our lives. However, nature has given us a vast repository, representing millions of years of evolution, wherein we can research, experiment and develop. A balance must be struck between cultivation of lands we truly need and leaving nature's "plant bank" alone. Of all errors made by man, destruction of natural plant life is his worst. Fortunately, the numbers of plant species endangered is not yet large. Man's negligence has been a symptom of his wantonness rather than an intentional crime.

A further reason for maintaining primitive plant life is to alleviate another serious issue, that of genetic erosion. We rely upon a very small number of plants. Experience indicates that resistance to disease is most pronounced when the variety of plants in use is high. We are also learning that the more complex the genetic structure, the greater the plant's capacity to resist disease. One gene, for example, produces stringlessness in green and wax beans. If a particular pathogen struck that gene, the vulnerability of stringless green and wax beans would become extraordinarily high. Since at the same time, the number of plants upon which humans are dependent is small, vulnerability of the entire agricultural system is high. We have become dependent upon a relatively few key plants.[6] In summary they are all vulnerable in fairly high degree to pathogens.

It falls to the husbandman, whether he be in field or laboratory, to recognize changes as these affect plants and to counteract these as effectively as he can. We must react sensibly. But in the modern world of specialists, it is common for someone to generate some sliver of information as to potential danger and to prophesy impending universal disaster. Linear extrapolation is used to "prove" the future disaster.

In the 18th century the white potato was introduced into Ireland from South America. The potato was new to Europe. In Ireland it found a potato heaven; it grew and proliferated splendidly. The potato yields among the highest caloric values per unit area of any plant under cultivation, comparable to rice and the cassava family. Its yield in calories is about 7 million per acre, which is similar to rice and three times that of wheat. The Irish population explosion may have resulted from ascendancy of the potato crop in Ireland. But the potato possessed a very narrow genetic diversity and in Ireland it was isolated from diseases to which it had built resistance. In 1830 disaster struck when a new fungus attacked the potato's narrow genetic base. Ireland still bears scars from the Great Famine that followed.

One could argue that the ultimate Irish disaster arose more from political action (or inaction) than it did from failure of the potato. The disaster does, however, leave us with two thoughts. First, it illustrates graphically the vulnerability of plants where genetic erosion occurs. Second, if a contemporary exponent of linear extrapolation had made such a projection in 1840 based upon Ireland, it could have been "proven" that in ten years the potato would have disappeared from the globe, as would all Irishmen. Yet today the potato flourishes — particularly in Ireland. The fact is that nature, and man as part of nature, does not always follow linear extrapolations. This fact is not well accepted by those who love to prophesy disaster.

Contrary to such primitive projections of disaster, plant genetics remains one of man's greatest hopes for increased production, even though it also holds dangers. This is what the stuff of life is made of, the challenge and how we respond. Toynbee put it well — we now face the challenge and the potential rewards promise us much. Rewards, however, cannot be forecast. When one endeavors to describe the future of anything that is important, inevitably one must fail. The degrees of superiority beyond what we have now are almost impossible to project as specifics. Regarding plant genetics, we can assume improvements will continue. But we do not really know. Neolithic man, crossing the Bering Strait to North America, could not possibly have foreseen modern America. Neither can we conceive future good and bad that might be done to our plants. What we can speculate on is promise. The promise held by plant development will at least match previous accomplishments. We must make a leap of faith that much can be attained, while we must also keep our husbandry at high pitch.

While plants are the basis of all human existence, animals also play a significant role in human sustenance. Animals live upon plants. Man consumes their flesh as a primary protein source. He uses their pelts for various clothing adjuncts. Parts of some animals are fed to other animals to produce more protein. Animal organs provide a staggering array of medicines and drug additives for humans. From late Neolithic times until the 1900's, animals were man's power source. Their muscle power was absolutely indispensable to agriculture. No creature on this planet has played such a role in the food chain for as long as the horse, donkey, ox and dog. It was a continuous role, patiently executed for about 12,000 years, perhaps not quite without breaks. For example, the horse collar was introduced into Europe from China in the 13th century, increasing average horse-pulling power some 40%. Man also found animals, even goats and dogs, indispensible to product transportation. Animal milk from cows, mares, goats, camels, sheep, reindeer, buffalo, yaks and llamas provided human nutrients in various forms. Raw milk, butter and cheese were the most common, but yoghurt, curds, whey, and exotica such as milk and blood drawn from the same animal must not be overlooked. For herding, man used animals, notably the horse and the dog. He used the dog to guard crops and herds.

Although it is not intrinsic to the food chain, animals were also companions to man. In American folklore, the cowboy talks to his horse during lonely days, and at night sings to the dogies. For cattle, especially younger cattle, are soothed by cadences whether it be of an imperfect cowboy voice or of a symphony orchestra rendering the "Pastorale" through a loudspeaker in the milking shed. Then there is the dog as a companion. How does one measure its service in the agricultural symbiosis? Besides, people who do not talk to dogs (and listen in turn), are unfinished humans.

Birds also have long been companions to man. Their song has awakened man in the morning and made him reflective at eventide. The flesh of birds was undoubtedly one of man's earliest foods, along with eggs. Their popularity has not diminished.

We cannot include fish as companions of man. But fish, including mollusks and crustaceans, have long been human foods. Shells of oysters, mussels, cockles and other species are found in ancient man's abandoned middens, as are fish bones and fish hooks.

Nevertheless, today meat and eggs provide little more than 5% of the food calories of the planet. These offer significant nutrients for those who consume them. Except for specific segments of humanity, fish plays a small role in human diet. Indeed rivers, lakes and oceans provide only slightly more than 1% of the world's food. In fact, man's relationship to the waters is still predatory, that of hunter and gatherer. There is little or no symbiotic relationship between man and sea, and such a relationship on a large scale is in the future, perhaps a not-too-distant future. The farming of mollusks, crustaceans and fish has begun. Americans, being quite wasteful, consume only one-third of the fish; the remainder is either thrown away or reduced to fertilizer or animal food. In contrast, Chinese and Japanese eat the entire fish. At a formal Chinese banquet the guest of honor is served the head, which is delicious, while the most insignificant participant gets the tail. All over Asia guts of the fish are eaten as delicacies. There is even a delicious Chinese tidbit made from the scales of one variety of fish.

Bird eggs offer the highest nutrition measured in essential amino acids and protein utilized. Next comes fish, then milk and milk derivatives. Meats follow closely behind. In terms of essential amino acids and protein utilization, vegetables do not meet the same standards as do milk products, fish and meat. However, a vegetarian diet can be wholly nutritious and esthetically pleasing to many people. But vegetables simply do not equal animal products in concentration of nutrients. This may be why animal products are in such demand, especially as people become more affluent.

At present some 19% of the planet's land surface is used as pasture land for domestic animal grazing. This compares with 10% of the planet's surface given over to crop production. Significant effort is now being made to supplement animal feeding by grains and other crops grown on cropping lands. As with plants and again in developed agricultural areas over this past century we can witness enormous increases in unitary animal products yields. This has occurred not because of increasing acreages devoted to pasture. These in fact have probably decreased. Rather it has come about by improvement of plant life together with improvement in animal genetics in conjunction with improved husbandry.

In many parts of the world we can note dramatic increases in animal yield — fleece, hide, meat and milk, but there are also other regions where there has been no such development. In the Sahel in West Africa, a native steer yields about 80 pounds of meat. To reach even this level, the animal could be from three to five years old, requiring that many years of food.[7] In the U.S. an average steer will produce 450 pounds of meat in two years or less...a matter of animal and plant genetics along with nutrition. The Sahel animal probably roamed over many acres of land to obtain its sustenance. For the last four to six months of its life, the American animal, fed on staples such as grain and cultivated alfalfa, drew its sustenance from less than an acre.

There are other examples. Native cattle in tropical areas not only yield poorly; they have low fertility rates. With inadequate nutrition and poor husbandry, nature protects animal life by limiting numbers of offspring conceived. Normally a bovine properly cared for in temperate zones has a heat cycle of about every twenty-one days. This heat cycle may last up to twenty-four hours. Tropical cattle have an irregular heat cycle and it may last for only a few hours.[8]

The calving percentage of a tropical animal will probably be below 20%. In the U.S. a rancher whose cow herd achieved below 80% calving rate would go out of business. One should note the significant early step increments that can be made with tropical cattle. Obviously if one can bring the calving percentage up from 20% to 40%, one doubles the number of calves born. This can almost always be done by improving husbandry:

by simple vitamin injections for the mother.

In 1977 the world produced 126 million tons of all meats, compared to a production of 1,434 million tons of all grains. Consumption of meats varies widely. In 1974 in the U.S., consumption of beef, veal, pork, lamb, chicken, and turkey amounted to approximately 237.6 pounds per person. Meats amounted to more than half of the American diet. Meat consumption in India supplied less than 0.3% of food calories; China 6%, Japan 5%, and SE Asia 3%. India, however, is a special case and will be considered later.

This imbalance in meat production and consumption arises from four interrelated reasons. First, in the planet as a whole, designation of land for grazing is small, and outside the U.S. there are few surpluses of grown foodstuffs to be used for animal food. Second, where land is available for grazing, such as the uplands of Southeast Asia, South America and Africa, the power equivalent for clearing forests has not been available, and there has been an absence of interest and motivation toward animal husbandry. Third, animals have historically been looked upon as work creatures rather than food. Fourth, where old animals have been used for food after their working life is over, the yields of flesh and its inferior flavor and texture have discouraged all but the most avid meat eaters.

In these countries we notice a phenomenon of which Japan provides the prime example. With the advent of general affluence, the first major change in dietary patterns is away from grains and vegetables, and towards animal protein. This trend seems to be universal. Man has a dietary need for protein; animal flesh meets that need. There are, however, many who object — not so much to eating animals as to production of this protein, especially in advanced Western countries, by feeding animals grains.

It has been asserted that animal conversion of grains to protein is wasteful. In China a person is adequately fed on 450 pounds of grain per annum. About 350 pounds are eaten directly by the human; the remaining 100 are fed to animals — chickens, ducks and especially hogs. In the U.S., however, average consumption by a human is closer to 1000 pounds of grain per annum. Only 150 pounds are consumed directly. The remainder, fed to animals, is consumed indirectly in the form of the 237.6 pounds of meats already noted. Should this be done? Should not these animals be limited, grain saved and distributed to a needy world? Would this not offer a more efficient agricultural symbiosis? It has been contended that it would, and an answer is deserved. Let us pause briefly to clear up some fallacies about animals and grain, particularly cattle.

It has been said that because a bovine requires eight or nine pounds of food to produce one pound of body weight, bovines are major culprits in grain waste. But bovines are ruminants. They need, for survival, large quantities of roughage daily. A bovine under American feeding practices consumes about 80% of its sustenance as roughage — that is, grasses, hay, silage, citrus skins and such like. So the bovine, more efficiently than most animals, consumes foods that cannot be used by man.

The pig, for example, needs only five pounds of nutriment to produce a pound of body weight. But the hog is a non-ruminant; it consumes a diet similar to that of man. This is one reason why hogs are so popular in many less-developed countries. The People's Republic of China (PRC) has about 250 million hogs, compared to 55 million in the U.S. In China this enormous production of hogs is traditional. Besides being very compatible in diet to man and in a general cleaning-up of his wastes, the hog carcass is sized for easy handling in societies without refrigeration. A 200 pound pig carcass is easier to handle and keep than a 1000 pound steer. To return to the point, in the U.S. a hog consumes about double the grain of a bovine to produce an equivalent amount of body weight, simply because a hog's digestive system cannot handle roughage as can his ruminant counterpart. Neither can man.

Next to the hog in order of grain consumption per pound of body weight would be chickens. Chickens produce a pound of body weight for every three pounds of nutriment. Today in the U.S. a chicken's diet is almost totally grain based. Cattle come third in this grain consumption derby.

Therefore if the U.S. were to "save" grain by not feeding it to domestic animals, the result would be something like this:

We still have large areas of grazing lands suitable for cattle. Possibly some 30 to 40 million animals could be supported upon these lands as opposed to approximately 135 million head supported by today's methods, a reduction of about 70% of the cattle herd. But the shortfall would be greater than that. It would take longer — perhaps four to five years, if ever — to produce a 1000 pound steer, as opposed to two years, today. Cumulatively the amount of edible meat per carcass would diminish probably by another 20% or more as compared to present averages. Prices would inevitably increase. Beef, and tough beef at that, would become something of a luxury. It is unlikely to happen.

If saving grain were the criterion, pork, ham, sausage and bacon would all but disappear unless individual families kept their own animals to consume wastes. That is impractical in urban America. The same condition applies to chickens and their eggs, which also are unlikely to disappear.

Rather than curtailing animal numbers, we should increase them through more efficient agriculture. Other animal species offer great efficiencies for man. The water buffalo, a remarkable animal, is probably limited to tropical regions because of its poor resistance to cold and its need for frequent access to water for temperature control; but it is in the tropics where the protein deficit is greatest.

Hitherto the water buffalo has been used primarily as a draft animal, eaten only when it was too old for further work. A steak from an ancient, work-worn buffalo is less a meal than it is a masticatory challenge.

Perhaps this is why the enormous potential of the water buffalo has not been exploited in most societies. He is a ruminant, of course, and a magnificent forager. His ability to convert food to body weight is unquestionably better than that of other bovines. Further, at a given age a water buffalo weighs about 50% more than an Angus or a Hereford. The cows are good milkers, giving up to four gallons of milk a day with 9-13% butterfat content, more than double the butterfat in Holstein milk. The younger animal, properly nutritioned and husbanded, produces tender, flavorful meat with much less fat (and hence waste) than a regular steer. The animal is more disease resistant for his geographical region than are other bovines. Other than for work, water buffaloes are practically an unused asset[9] because of cultural restraints.

Without invoking dramatic research or development programs, room exists for quite important improvement of animal production by simple means. In fact, this situation is common to virtually all tropical agriculture.

In many respects this discourse on animals and their grain consumption has been a diversion, but a deliberate one. It illustrates many facets of the symbiosis between man and nature in man's effort to produce food, clothing, and shelter. If we are to effect symbiosis we must understand in intellectual and practical terms that agriculture is a web of relationships. Agriculture must be understood not only in its specialties but as a relationship between specialties. So far we have looked at but a few elements. These — husbandry, plants and animals — are essentially practical. We have yet some more elements equally practical to explore: soil...climate...work, the latter an extension of the husbandry concept.

We are also beginning to bring in our first seemingly non-agricultural question. Feeding of grains to animals when humans need grain, and the concept that producing animal protein from grain is "wasteful" intrudes social and philosophical issues. What is the right ethical, if not economic, relationship between those who produce agricultural abundance and those who do not? If a free society must give away its abundance, then under what condition should this be done; or, are some of us imagining a less free society where one is forced to give? Above all, hanging over us we have the same nagging question. Why do these enormous quantitative and qualitative imbalances in world food production exist in the first place?

Perhaps before looking at plants and animals we should have looked at soil. Soil takes on an almost mystical quality in some cultures, and tends to be regarded as the basis, even the guarantor, of life itself. We talk of the blessedness of rich soil and deprivations born of poor soil. Many long to own it; sometimes have a passion to own it.[10]

But the significance of soil *per se* may be inconsequential. Soil may be little more than that substance which provides a holding structure for roots of plants and for the moisture and nutriments plants need. As a rule of thumb, an adequate soil is one that can carry moisture (and thereby nutrients) to some 40% of its weight. It is these other elements, especially water, nutrients and sunlight, which have real significance in terms of plant life. Virgin soils sometimes may be "rich" soils, with large reserves of nutrients. Continuous cropping, however, will soon deplete these nutrients and rich soil will become poor soil.

In 1840 a scientist named Liebig enunciated the "law of the minimum" which states that the yield to be obtained from any crop is limited by the amount of any one of the minerals existing in a particular soil as essential to plant life. Thus, those essential minerals have to be replaced as they are used up by plants if continuous cropping is to be maintained. Fertilization is essential and cannot be escaped. On the other hand, we have concentrated deposits of phosphates, potassium, lime and other minerals essential to prolific plant life. It makes sense to mine these deposits and sow the result in the right amounts in croplands and pastures. There is little doubt that the original virgin or "rich" soils have been made richer where this has been done in accordance with good husbandry.

Animal manure, humus and other organic material from plant residues also help build structures in the soil which carry essential nutrients. But inorganic fertilizers can be added more efficiently than organic materials. Soil carries the plant's root structure; if soil also contains the appropriate nutrients, it allows the plant to feed and grow. If nutrients are not in the soil, they must be put there. If nutrients are in the soil, they must be replaced as plants consume them.

Plants can be grown without soil. Hydroponic farming consists of mixing the right nutrients with water, which is absorbed by the plant's root structure without soil. In ancient times farmers in arid regions of China sometimes used this method in a water-short region. They condensed water at night through the medium of pebbles placed in an inverted pyramid where differential rates of cooling promoted condensation and the inverted pyramid produced drops of water. What nutrient was mixed with the water is unknown, but it probably was human or animal feces.

Significant hydroponic experiments are being conducted in many areas. In Abu Dhabi[11] beach sand is being used as the holding structure for plants. They are then fed water from the sea, de-salinated and loaded with nutrients. Hydroponic farming, other than for specialty crops, cannot yet compete economically with regular cropping, but technique illustrates the flexibility of plant life. It opens up another dimension in our agricultural future, for if and when hydroponic farming becomes viable economically there would be virtually no limits to the amount of food that might be grown.

Despite a plant's flexibility, it will not grow in all soils. Indeed plants have preferences. Tolerances of one plant for, say, an alkaline soil will not be matched by another. There are substantial parts of the earth's surface where plants will not grow at all because of unfavorable mineral conditions. There are irrigated regions with an alien buildup of unfavorable minerals caused by pumping mineral-laden underground water. If not handled properly, these mineral buildups can make soil infertile.

In the very worst cases a difficult decision arises. Should one grow crops on this acreage for 100 years knowing that the soil will eventually be rendered useless and need expensive rehabilitation? Or does one make a leap of faith and assume that by man-induced changes in genetic structure, the plants will become resistant to these harmful mineral deposits, or that new water sources will be found; or that new methods will be found for treating soil? Here is our familiar trade-off again. Here are some of the hard questions that arise continuously in practical agriculture.

Perhaps the most important point is that there are few naturally "rich" soils and that richness will soon be expended by cropping. Soil requires continuous care; it requires good husbandry if it is to produce continually. One senses this in a Japanese rice field. It is a tiny square barely large enough to allow turning room for a small-sized, hand-held agricultural tractor. But the field has been producing rice, vegetables, perhaps perennial fruits, for 1000 years — or longer. It will produce for another thousand years — or longer, because its plant nutrients are periodically renewed.

Man is the architect, a peasant with bare legs. His feet are oblong to match the symmetry of his field. His head is domed to match the heavens above him. The blood courses through his muscular legs and his lean body, and his knowing brain and flesh link heaven and soil. It is to him we pay tribute. Joined not only to soil and elements, he is also joined to ancestors who husbanded this same soil. It has been they and now he who sustain human life, and life yet to come.

If soil is the holding agent for plants, climate is the stimulator and often the vitiator. The two most significant elements in climate are water and temperature. Where there is no water, there are no plants. Where there are no plants, there are no animals. Of all substances upon which plants depend, water is the most fundamental. The chemical reactions which take place in plant cells to allow burgeoning life are stimulated by water. Any substance which might be needed throughout the plant's system can be moved there only by water. Without water, plant metabolism is dormant, and usually, plants die.

Seeds of a plant, that part which contain the life forces and the code which determines the plant's configuration, can lie inert for years... certainly hundreds of years, perhaps thousands if they are kept perfectly dry. This in itself is invaluable to man. It gives him, as husbandman, enormous flexibility as to how and when he might utilize that seed, including ability to transfer the seed and its life force over time and space.

Most seeds can withstand freezing temperatures and great heat. No matter the length of dormancy or temperature extremes, when that same seed is given water it springs to life. In general when a seed absorbs water equal to 8% of its bulk, the process of respiration begins. Water penetrates its husk in molecules. When the water absorbed accumulates to 12% or more of the seed's bulk, germination begins and the plant starts growing in accordance with its genetic code.

It is mandatory that water is applied in the right degree. If insufficient water is applied, (less than 12%) the seed will not germinate and instead will become infertile. Too much water may cause the seed to rot. After germination the embryonic plant rapidly develops a root structure, which draws water from the soil by osmosis (absorption of a liquid through the cell's skin). Thereafter a complex series of tubes distributes throughout the plant's system moisture-carrying substances essential to growth. Plants *do* have one distinct limitation in their utilization of water. Water must come to them; they cannot move to water.

Within limits, however, animals can move to the water they need. A cow ought to drink every day. Other animals may obtain adequate water from their consumption of plants. Because of this mobility, animals have found a more viable way of meeting their moisture needs than have plants. Today domestic animals have water provided as an essential part of good husbandry. One of the more satisfying aspects of cattle-raising is to watch mother cows at about 4 p.m. each summer day begin their trek to the water hole. A single cow may be left behind to look after young unweaned calves. The cowboy will watch the process carefully, making sure that there has not been a lapse on the part of some animal in misjudging distances in the grazing pasture and perhaps straying too far from the tank. This is rare, for animals judge their day well and the water hole is focal point in this judgment.

By far the greatest amount of water, especially for plants, comes from rainfall. The single most important agricultural resource the planet possesses is rain. Temperature, soil constituents, plant type and sunlight are all critical factors, but rain is the determinant. As the planet's surface bears witness, without rain we have deserts. With limited rain we shall have only hardy plants and trees such as cactus. In medium rainfall areas there are grasses in a mantle covering everything. Heavy rainfall regions enjoy forests. At the lighter end of the heavy rainfall spectrum there may be birch, beech, ash, oaks and conifers. At the other end we find the incredible diversity of the tropical rain forests. (In the latter case, however, temperature also plays a key role.)

Temperature is important. Some plants have extremely small tolerances to temperature differences; others more so. Along with rainfall, temperature is yet another key determinant as to what plant will flourish and where. The tomato, for example, will have its fruit "set" when the temperature is at about 65°F. This is why a tomato crop in a given location can vary from year to year by as much as 50%. Other than growing tomatoes under an enclosure, there is little that the farmer can do — except once again to doff his cap to nature. All plants react to temperature. Temperature in fact determines the territoriality of plants.

Sunlight is the key factor in the climate element of symbiosis — regulating temperature, affecting rainfall, and being directly related to photosynthesis. Photosynthesis is the determinant of the way a plant grows. The sun is the arbiter of climate *per se*. The entire meteorological phenomenon of moisture patterns, temperature patterns and seasons is directly related to the sun. Beyond this, plants themselves have a quite intimate relationship with the sun. For example, each leaf of a plant grows in a spiral pattern along the plant's stem. That is, if the spiral from the topmost leaf is counted downward to the next uppermost leaf and so on down the stem, it will be found that in accordance with the functional formula no leaf shades the one under it from its share of sunlight. There are some exceptions in nature, but generally this holds good for all common plants.

Air, that other key constituent of the biosphere, must also be added to climate in our symbiotic lexicon. Plants must draw carbon dioxide through their leaves for photosynthesis to occur. As part of that intricate process carbon dioxide is converted into carbohydrates acting through the chlorophyll of the plant under light. Suffice it to say that here water, temperature, sunlight and the use of air by plants are intrinsically related and interdependent.

When one views the impact of climate upon plants, ofttimes one's reaction is one of helplessness in the face of such immutability. Surely man can do little here. Surely all doctrines of good husbandry must be dependent upon climate's whim. This is not entirely so. Certainly, massive changes in weather patterns or cycles could (and often have) completely changed agricultural patterns. About twelve thousand years ago we came to the end of one ice age. There will be another — but when? What would the construction of American agriculture be with areas such as Illinois under an ice cap of perhaps one mile in thickness? Would the now arid southwest then flourish under a new-found abundance of water? Obviously with an ice age, new relationships in American agriculture and in planetary agriculture would arise. Our capabilities to adjust as humans will be tested. Then there are less dramatic changes such as droughts, floods and temperature variations.

Farming, however, must continue as of tomorrow and the next day. Ice ages will have to be faced if and when they come, as would equally dramatic circumstances should a warming trend occur. Droughts and floods have to be managed as best we can in the shorter term. As husbandmen we are tested. One part of the lot of humans is and always will be to react to shortfalls in nature, in terms of man's arbitrarily stated needs. Man has shown in a decisive way how to react to one climatic impediment: irrigation is one such reaction, and without doubt the most important.

We know that by 3500 B.C., Egyptians were irrigating the Nile Valley and by 3000 B.C. Sumerians, the Mesopotamian Plain. By 2000 B.C., at the latest, the Chinese were irrigating the Yellow River Valley. In North America the Hohokam Indians were irrigating parts of what is now Arizona, using alignments almost identical to that used in the 20th century by Europeans. Irrigation spread, but not nearly as extensively as it could have. Part of the theme of this book is to explore this aberration; why man does not always utilize resources about him.

Early irrigation was basically fairly simple. It consisted of transferring water from a river, lake or pond to adjacent fields. In some instances canals were built. Sometimes these served a double purpose of irrigation and transportation. Another method was to capture water behind dykes for later use. Full use of rivers, however, was not developed until modern engineering gave us the capability to build large-scale dams. In the twentieth century irrigation changed dramatically, as modern engineering made possible the damming of large rivers with resulting lakes of enormous sizes. Modern earth-moving equipment and the use of cement allowed for reasonably economical construction of large canal systems, which in turn allowed for the distribution of water over large areas.[12]

One of the most significant developments in this century has been pump irrigation. There are vast deposits of underground water in many parts of the planet, often under seemingly arid regions. Some of these aquifers are static in the sense that under current climatic conditions they are not recharging. Thus every gallon of water drawn from a static aquifer is a net loss; water is being mined, so to speak. Other aquifers are constantly recharging naturally. Here, under favorable conditions, a balance can be achieved between what comes in by natural means and what is taken out by man.

Irrigation from underground water sources and from dams is more significant than it first appears. For the most part, under natural conditions, the provision of essential water to plants by rainfall is full of surprises. Rain comes at unspecified times and in unspecified amounts. In many climatic regions this natural system works well. In others it does not.

Where aquifers are recharged or are large, or there are dams fed by rivers, the agriculturist could have a most favorable potential ahead of him in balancing out weather phenomena. Indeed, beyond a precisely controlled precipitation pattern (which does not occur), the most favorable agricultural situation would be for rain or snow to fall in nonfarming areas, for resulting water to drain and be stored underground and then pumped out on a demand basis for use on farmlands that are perpetually basking in sunlight. In limited degree this is precisely what is happening in the southwestern U.S. and in a few other regions of the planet. A juxtaposition of stored fresh water and warm desert gives us one of our most productive agricultural prospects.

There are other areas where water is too plentiful. It can fall in immense quantities over short periods. This is of little value to agriculture except where a water control system is in operation. In fact, flooding, often on a disastrous scale, is the lot of many regions normally water-deficient. There are also great areas that are (or were) water-logged, where the problem is how to get rid of too much water. In regions such as the Amazon Basin and Southeast Asia there can be too much or too little water at different times of year. Water remains a vital key to successful agriculture. But water is haphazardly distributed throughout our planet. Philosophically, perhaps this is either a great accident of the Universe or has been purposely so designed to test man. Accident or design, the challenge is there.

Although man has shown great ingenuity in controlling and redistributing water, he has barely scratched the surface. The challenge, with its opportunity and undoubted reward, is still ahead of us. Irrigation is the most potent single element in agriculture today and for tomorrow. Desert land allied with water offers us our biggest opportunity. Except for the high cold deserts, desert temperatures are such that selected crops, in surprisingly wide variety, can be grown for all or most of the year. Continuous sunshine promotes continuous photosynthesis. Planting and harvests are not interrupted by rain or snow. Work can go on continuously. Today it is common in the Southwest U.S. for one field to produce up to three bales of cotton per acre (about twice the normal U.S. yield) and then a grain crop which matches or betters national averages, all within one year.[13]

The world's arid regions offer a tantalizing opportunity. With water they may become the most productive agricultural regions on earth, but how to get water? It will not come from nature short of another dramatic, unforeseen and unlikely immediate change in the world's climate. Man has to do the job if it can be done. De-salinization of sea water is still too expensive. Although intensive efforts are being made to improve this, that could take along time. We have plenty of fresh water in snows, rains, rivers, and aquifers. We have to redistribute these resources. Yet a precise, comprehensive planetary water map has yet to be made. Yet water redistribution under the aegis of man is an essential element in promoting agricultural plenty. We have shown admirably what we can do on a limited scale. We now have to expand our efforts. Here is an exciting challenge, one geared to construction rather than to mankind's more common challenges geared to destruction.

We should, however, note one further issue. We return again to a trade-off. Underground water in particular is often much more saline than precipitated moisture, and contains other adverse minerals. While this mineral content may not adversely affect a given crop, continuous application over a long period of time can build up mineral accumulations in soil which make it sterile or partly so. This has happened and can occur again on ever widening scales as irrigated agriculture increases its dimensions. Techniques, principally leaching the minerals from soils by additional irrigation, is normally effective. But the danger is there as yet another element in this symbiotic relationship between man and nature, as man pursues his quest for food.

In symbiosis, as we build it up, we have considered plants, animals, soil and climate, and the relationship of each with the others and with man. We must now bring man even more to the fore. Let us look at work, another word for husbandry.

Work comes in a variety of forms, but short of the extraordinary limitations of a hunting and gathering society, no food or fiber is produced on this planet without some expenditure of human energy in terms of both muscle and brain. We shall examine this activity in more detail later. Here we shall view it as part of the symbiosis. That farmers work is well known. That the work is often hard, long, enduring and in the face of all weathers is perhaps one reason why persons the world over have sought to leave the farm for urbanized living. For some farmers, in the U.S. for example, the trend has been away from muscle power and toward brain power. This trend must and will continue for everyone. But we would be deluding ourselves if we were to assume that mechanical energy will replace human energy. For many a farmer, his daily work will remain physical as he, with his body, accomplishes tasks that machines may do elsewhere. As of now and for any foreseeable future, we must face the fact that the bulk of the planet's farmers will be engaged in physical toil. Can it be made more rewarding toil? Yes, it can.

What is the objective of a farmer's work? It varies from culture to culture. Some African herdsmen desire larger numbers of cattle rather than better cattle and end up with a few more but very poor cattle. A peasant's measure of success may be enough rice for his family for the ensuing year. An American farmer will look at yield, how much it cost him to obtain that yield and what he can get for it. Nevertheless, farming people all over the world who sell all or part of their crop have the same objectives: "How much can I grow? What does it cost? Where do I take it? How much will I get for it?"[14]

The best goal for agricultural work is obtaining maximum yields and for these, obtaining the best prices at the farm. As ever there are trade-offs. Sometimes it is better for a farmer to take a lesser yield if the cost is relatively lower than that of obtaining a high yield. By and large, however, maximum yield is the best measure of agricultural effectiveness. It also seems to produce a solid sense of accomplishment — a psychic reward, if you like — for the farmer.

It is suggested that we could adopt world-wide goals for increased yields and that these might be a measure of work. We should strive to double in a decade gross production of food, fiber and animals. First, we need to double the farmed areas, principally in South America and Africa. South America would then farm some 24% of its potentially arable lands and Africa some 30%, representing quite vast areas of farmlands.

The next and cumulative goal should be to double yields per unit of area farmed in South America, Africa and in Asia, excluding Japan and Taiwan. For the most part yields in these regions are so low (except in

Japan and Taiwan) that doubling them is theoretically a not very formidable task. Such increases as outlined would about double the planet's food supply. We shall explore how these goals can be accomplished and how they are impeded at present. Further, if past history is any guide, a modest two-fold increase would have an even more important stimulatory effect. When one begins from a low base, increased production tends to beget further increases. We might then become serious about agriculture. We might begin to envisage those tenfold increases, or greater, which are inherently possible as one result of proper work.

Obtaining yields as part of this man-and-nature symbiosis requires work; not merely raw work, but work of a particular kind. Agricultural work is skilled work calling for good judgment as to plants, animals, soil and climate. Mistakes can be very costly. In subsistence agriculture, a mistake can mean death. In advanced agricultural technologies, a mistake will always cost money, probably affect other ongoing and future agricultural operations and cause some loss of one's self-regard. Judgment is the key factor of sound farm work. From preplowing through plowing, sowing, cultivating, harvesting and marketing, a farmer is faced with an interweaving web of judgments.

There are yet other relationships — those of relating agriculture to other parts of our economy. External relationships occur in any agricultural economy where surpluses are marketed. Agriculture is America's single largest business. First comes growth of the agricultural products as they provide approximately 2,500 calories daily for every American, and enormous surpluses besides. Allied with that is the continuing need for clothing and shelter, both of which have to be constantly renewed. Then comes building and maintenance of farm and agriculturally related infrastructure. Machinery might be next, not merely farm machinery, but also motor vehicles and transportation, product handling equipment, specialized irrigation equipment, offices and computing and accounting machines, electrical equipment, scientific and research equipment and the special machines to make equipment.

The chemical and mining industries are very much part of agriculture, providing fertilizers, pesticides and special medicinal products for animals. Agriculture requires energy, electrical energy, gas, to move water and provide sanitation, to sow and harvest. Packaging of agricultural products is a major supportive industry. Packaging prevents spoilage and depredation by insects and fungi, makes for easier handling and more efficient transportation. The list of industries involved in packaging is long. It includes textiles and floor coverings, plastics and other synthetics, rubber, glass and metal containers, lumber and wood products and paper products.

One of the largest supporting industries in terms of cost is that of various services on behalf of agriculture. These may include: office supplies, printing and publishing, warehousing, shipping, trucking and railroading, retailing, real estate, hotels and other personal services, repair of agricultural equipment and personal equipment, scientific research and development, accounting, banking and other financial services.

This, then, in descriptive terms, is the work interrelationship of agriculture in the U.S. To a lesser degree this applies to virtually all agricultural systems in the world. Twenty years ago a Thai peasant wrapped his food products in a banana leaf. Today the product is wrapped in polyethylene. There was formerly an industry in gathering of banana leaves; but no more. It has been supplanted by the manufacturing of polyethylene bags.

This interrelationship of work to work, and work over all to the rest of the symbiosis is perhaps the most complex aspect of agriculture except the workings of nature itself. All stems from nature, from plants, animals, soil and climate, with man as the stimulator. The question is one of symbiosis, of each element relating to the other, contributing to the other's support. When we look further at the symbiosis we see that it is there as a basic condition. It must exist if there is to be agricultural productivity and for all those other things that spring from such productivity. But there are variances in the symbiosis that are extraordinarily important.

Why is the rice yield in Japan at 2.4 metric tons per acre so much greater than that of most other Asian countries? We see Laos, with .56 metric tons per acre or India at .72 metric tons per acre. Why has the rice yield in Japan increased from the .4 metric tons per acre recorded in 1850, to that of the highest yield in the world? Here is but one example of yield imbalance. There are others.

Why is only 22% of Africa's arable land farmed? Why, at the same time, must the hunger occurring from the Sahel's drought be alleviated by grain from Europe and America?

Why will farmers in some monsoon rice-growing areas do nothing after the monsoon-inspired crop is harvested until the next monsoon, a period of about eight months in the year? More importantly, why will other farmers in monsoon areas continue to work year-around?

In 1972 this writer was having lunch on a two-and-a-half acre farm in Taiwan. The farm supported three families on a scale of material luxury undreamed of in other peasant cultures: color TV, a refrigerator, and a washing machine (who could imagine a washing machine in the home of a Chinese peasant), several sewing machines and several Hondas. During lunch I asked why they worked all year around when one could fly in a mere six hours to a southeast Asian country where, once the 4-month monsoon rice cycle was over, the farmer did nothing for the remaining eight months? An old lady at the far end of the table became quite disturbed at these remarks. Later I asked why. The farmer laughed. He acknowledged that she was quite upset. In fact she said she knew the foreigner was a fool the moment he came on the farm. She added that now she knew he was a liar, too, because no peasant would just do *nothing* for eight months! But some do, and we must ask why.

Why is one depressed by the apathy of an Indian rural village while at the same time immigrants from the same Indian village become smart businessmen in Singapore or Hongkong? More dramatically, why do Indians in the islands of Fiji become highly enterprising farmers?

Why could Burma export three million tons of rice in 1940 and by 1960 be exporting virtually none? Why can one find that in lands where people are most hungry, rodents and insects eat better off the scanty harvests than do the people?

Why is it that of the 3,360 million acres under cultivation, only an infinitesimal fraction grows more than one crop per year? Why are the yields of crops generally so low, except again for a tiny fraction of the planet's farmed area? Why is it that approximately 3,080 million acres of potentially farmable land are not cropped at all? Why have we not bothered to establish how much arable land we might have under optimum conditions, for it is surely greater than the estimates given above?

The rhetoric is disturbing — much more disturbing than rhetoric ought to be. The questions asked are but the tip of the iceberg. The list of human failures in contemporary agriculture is near endless. The list of successes, though impressive, is small. The greatest success in agriculture has been symbiosis itself. Yet symbiosis only works to its full here and there. In terms of what might be achieved through better relationships between man and nature, we have barely begun.

We cannot blame nature. True, as we have noted, nature is far from perfect as we would have perfection. Climate is erratic and benefits randomly distributed. Soils offer as many impediments as they do impellents. Why should not a cow give twice as much milk, or an ear of wheat twice as many grains? These are the constant queries that must be made by a farmer as he confronts nature. We might also observe that nature is a hard taskmaster. To succeed, one must do it right. Mistakes are not forgiven. Nature is cruel, arbitrary and indifferent to us. If one does not accept this, twenty-four hours exposed to the Arizona desert or a Minnesota winter will make the point. But nature is a giver. We must begin by taking nature as it is. Our man-induced changes to nature, those which we must make to survive, are changes to be made by respectful relationships rather than through human arbitrariness. Whatever else we do, we cannot, as yet, blame nature for our shortcomings. It is man who has failed and is failing. It is man who will then fail to adjust, change and adapt nature through work to obtain a yield of food and fiber to meet his needs. Thus, if this book has a central theme it is that of man rather than agriculture. Nature is there, a bystander, willing to play a role if approached properly.

FOOTNOTES

Unless otherwise noted, general statistics in this chapter were compiled from data available at the National Agricultural Library, U.S. Department of Agriculture, Beltsville, Maryland.

1 U.S. *Congressional Record,* "The Economics of Food in American Foreign Policy," 94th Congress (Senate), Sept. 1975, S. 15110.

2 *Scientific American,* Sept. 1976 (entire issue).

3 *Ibid.*

4 G. Borgstrom, *The Hungry Planet* (New York, The MacMillan Co., 1965), p. 5.

5 It is generally agreed that males have a higher requirement than females except when the latter are lactating.

6 It is undoubtedly true that one advantage of hunting and gathering was that the gatherer had a considerably greater variety of plants and plant ancillaries in his diet than is the case with modern man.

7 Victor D. DuBois, Food Supply in Mali, Vol. XVI, No. 1 of *Field Staff Reports, West Africa,* American University Field Staff, Hanover, N.H., 1975, p. 6.

8 This is why artificial insemination works poorly in the tropics, especially under open grazing conditions.

9 The configuration whereby small peasant owners, owning one or two water buffalo, could realize a beef industry, lies outside the concerns of this essay. There are many ways in which this could be achieved.

10 Cf. the story of how one peasant, offered ownership of all the land he could walk around in one day, broke his body trying to walk around too much of it, and ended up with only enough in which to be buried. Leo Tolstoy, *How Much Land Does a Man Need.* From "Russian Stories and Legends," Trans. Louise and Ayrmer Maude, (New York, Pantheon, 1967).

11 In the U.S., the parent of this project is the Environmental Research Laboratory, University of Arizona, Tucson, Arizona, under the direction of Carl N. Hodges.

12 The Central Arizona Project, due to distribute its first water in Southern Arizona in 1985 will have a canal system 307 miles in length. Nevertheless in Thailand a canal system exists of more than 500 miles which was entirely dug by hand in the 18th and 19th centuries. The Thai system was not, however, constructed with irrigation in view, but to develop an all-weather transportation system for military purposes.

13 Arizona is now the largest producer of Durum wheat, a "hard" wheat, in the U.S. This wheat produces specialized flours. Formerly the North Central and Middle West, dramatically different from Arizona, held this honor.

14 Ronald C. Nairn, *UN Aid to Thailand: The New Colonialism* (New Haven, Yale University Press, 1966).

3 AN IRASCIBLE CREW

Why are so many people not being fed, clothed, and housed adequately? The most commonly given reason is over-population; consequently, the solution to the dilemma should be to reduce population. This is a double fallacy. The over-population issue is part of a wider concern, insufficiency or waste of natural resources. Obviously if over-population and resource deficiency are taken at face value, the two are intrinsically related and ought not to be separated.

Those who most fervently propound both dilemmas are the members of the environmental movement. Because so many of the tenets of this movement fall within the framework of the over-population and insufficiency-of-resources syndrome, the approach here will be to look at the environmental movement as a whole and how it might relate to agriculture, especially in the future. This is a diversion from the main theme of this book — how to grow more food. The themes of over-population theorists and environmentalists are now part of our times, and without doubt influence thought patterns of our world. It is further posited that "how we think is how we grow things"; our agricultural actions arise from our thoughts.

One characteristic of the twentieth century, and especially that of intellectuals, is a strong antipathy towards the economic dynamism that has been typical of agriculture in the West these past two centuries. This turning away has reached its coherent form in recent times in ecological or environmental movements. These movements are not yet an impediment to agriculture. After all, if a farmer is not an ecologist, who is? If a farmer does not practice conservation, he is not a farmer for very long. But while the doctrines of contemporary ecologists do not in the main attack agriculture, their pronouncements do raise challenges to agriculture.

One pronouncement, always stated categorically, has for many become an article of faith: that for both short and long term there are too many people relative to the planet's capacity to feed, clothe, and shelter them. People, in this view, have become liabilities instead of assets. This argument can be very persuasive, backed by supporting data and often a good deal of passion.

A second pronouncement is that the planet is running out of finite resources and that this alone calls for drastic restructuring of human life patterns, including an economic restructuring. Agriculture, especially modern agriculture, being a symbiosis embracing many parts, obviously must be drastically affected materially and psychically by the disappearance of resources which cannot be replaced. Again much passion surrounds this aspect of the ecological movement. Draconian measures are advocated to "save" resources.

The third plank of the movement concerns pollution. Pollution seems to arouse people most. This is probably because they can see it. Pollution, it is said, arises from the profligate use of resources and is symptomatic of the dissipation of those resources, which is bringing us to ruin. Even more dire things have been said about pollution in terms of not being able to breathe the air, and drink or make other uses of water. Most agriculturists tend to ignore pollution, which, except for isolated or quite specific instances, is not a part of their modus operandi. It must be dealt with here because it is so much a part of the movement, which in fact challenges premises intrinsic to agriculture.

For example, a curtailment of energy production because of pollution curbs, affects agriculture directly. A similar situation arises from curbs on raw materials production. In effect, this means less and more expensive food. Agriculture therefore could be pushed into the invidious position of being denied materials to do the job, therefore failing to feed larger numbers of people. It would be claimed that this proves ecologists correct: there *are* too many people and we *are* running out of finite resources. Thus agriculturists must take note of what ecologists are saying and doing.

There is a fourth element in the ecological movement. Although lacking a substantiating body of doctrine, it has sufficient zeal and certainty of belief to allow one to discern certain religious overtones, albeit secular ones. Unless the ecological movement intends to become a proselytizing movement, this in itself is the business of only those concerned. It is their right. But like any other religious movement under the doctrine of separation of church and state, the movement must be denied the right to force its doctrine on others by legislative fiat.

Again this is not an uncommon nor necessarily a parlous state in the human condition, and one which usually rights itself through a rational reconciliation of inconsistencies. Ecologists are now proselytizers. They are asking governments to take action that would seriously impede productivity, which in turn would directly affect agriculture. More importantly, they are impinging upon people's confidence and this is truly serious.

The present population of the planet is about four billions and it is increasing. There are claims that we are in a population "explosion" which is bringing disaster to the planet and which will bring even greater disaster. The logic states that current food deficiencies will get worse and worse, resulting in mass starvation. Beyond food, the logic states, nonrenewable resources such as minerals, oil and gas will decline in availability even more rapidly than food. Even fresh water, we are told, is becoming a finite resource, unavailable for farm expansion and in many cases polluted beyond recovery. As a capstone, the logic says that an indefinable "quality of life," is deserting us. Fortunately, these dire prognostications come almost entirely from intellectuals, and history indicates that we should be overly careful about taking too much heed of intellectuals as either realistic analyzers or as prophets.

Intellectuals are prone to crisis talk and thereafter to preachments as to how they can save humanity from crisis. Sometimes intellectuals have been known to create crisis configurations leading to self-fulfilling prophecies. These characteristics are very prevalent in the contemporary West.

A particular reason for crisis configurations by intellectuals arises from their attraction to the misleading art of linear extrapolation. By using selective data and projecting these in a straight line, intellectuals can develop gloom and doom forecasts.

The study by the Club of Rome, "Limits to Growth," abused linear extrapolation the most. The study attracted great attention: it stimulated some hysteria with its gloomy prognosis. Fortunately the errors of the study have been revealed and it is now passing to the oblivion it deserves.

World population expansion in the 1950's could be projected to increase ultimately at the speed of light. As one tongue-in-cheek commentator pointed out however, that situation would never occur. Heat generation of closely packed bodies would create mass deaths. This is an extreme case of linear extrapolation, but other equally ridiculous forecasts can be made. If our intellectual friends had been in action at time of the Treaty of Westphalia in 1648, they could have developed an even more disturbing forecast. Europe had just suffered its most devastating of wars. More than a third of the people of what we would now call Western Europe had died from war, related famine and disease. In linear terms, in a few short years obviously there ought to have been no people left. Humanity, however, does not move like that. Rates of population growth change. They do not inevitably over time follow straight line extrapolation.

Instead, population moves in cycles. It increases, then decreases; moves upwards, then downwards. It seems that these movements are tied to economic growth. In a recent study dealing with Japan it was noted that:[1]

> **When industrialization reached a certain level (i.e. at the early stages of economic growth) the rate of birth did rise rapidly. We, the Japanese, recall that this population explosion of Japan was often regarded at that time by the West as a serious threat...The Yellow Peril...as soon as industrialization went beyond a certain level the rate of birth began to decline...as the rate of economic growth rose, the rate of reduction of the birth rate also increased ...The case of Japan tells us that sooner or later economic growth is bound to reduce the rate of population growth to zero or even to some negative rate, and that in return this reduction of the rate of population growth is bound to reduce the rate of economic growth. It is very impressive that both economic and population growth have such self-correcting mechanisms which will check their own exponential explosions.**

This thesis has been expounded many times and in other contexts. In the United States, for example, the rate of population growth in 1800 to 1810 was 3.1%. In the decade 1960-1970, the growth rate was 1.2%. All indications are that the growth rate will probably fall below replacement levels in the coming decade.

In a recent United States work, reasons for population decline were phrased more in sociological than in economic terms as was the case with the Japanese version, where, incidentally, the rate of population growth is now 0.6%. Sociological reasons given were as follows:[2]

> **• The abolition of work and the establishment of compulsory education for children, which results in increased child-rearing costs, and the delay if not the elimination of the time when children can begin to contribute to the support of the family.**
>
> **• Modern child-rearing practices, which require more parental attention. The child used to bring himself up, as Sauvy says, but now he is 'the center of much attention. He has acquired more value and more importance and his birth has become kind of an event, about which people think twice.' In addition, it could be added that parents who follow 'democratic' practices of family management might think twice about having more than two children in order not to be 'outvoted' at family council meetings.**
>
> **• The emancipation of women, which results in an increasing variety of non-maternal interests and careers for women to pursue.**
>
> **• Opportunities for social advancement, which create an incentive for parents to have smaller families so they can better prepare their children to take advantage of the opportunities.**
>
> **• The development of education, increase in the living standard, greater consumption possibilities, and growth of cities, all of which have the effect of tremendously broadening people's interests and activities, thus resulting in a decreasing emphasis on child-bearing as a source of satisfaction.[2]**

In the underdeveloped world, the contentions given above are more or less confirmed by reverse process. Children are cheap, not expensive. In a small village on the island of Luzon in the Philippines, one family, when asked why they had so many children replied, why shouldn't they have children; they cost less to keep than a dog and were much more useful.

In a small village near Lamphun in northern Thailand, years later, I asked the same question. The response, in an earthy vernacular, was that the village had no electric light, there was nothing to do at night, so each villager copulated "every night."

In underdeveloped country after country families view the value of children as workers in the field and as security when parents grew old. With these attitudes, children are a virtue; they are assets, not liabilities. For smaller families to eventuate, such attitudes have to change.

Basically, human fertility is a function of moderation rather than contraception and other fertility controls. If motivation is lacking, no amount of contraception technology will have any effect on fertility. Conversely, if motivation is present people will find a way to reduce fertility regardless of contraceptive availability.[3]

Thus once again we come upon the issue of human motivation and will, rather than technology or technique, as the prime factor of development, whatever form that development might take as long as it results in economic enhancement. A sophisticated agriculture simply does not call for large families.This applies to sophisticated small-scale agriculture such as might be conducted by a new style peasant equally as it does to large-scale heavily mechanized agriculture. The advent of these sophistications will cause a change in human attitudes towards families. History would indicate that as affluence increases family size decreases. Thus declining population rates are another outcome of vigorous agricultural development. Such should confidently be expected.

Perhaps the greatest fallacy is that population reduction in a poor country will solve that country's food and fiber problem. Agricultural production in the traditional mode is geared to human labor. Thus if you reduce the availability of human labor among the farming fraternity at some stage, there will be a corresponding fall in productivity. This is not entirely true because there is large-scale underemployment in most countries clinging to traditional agriculture. Population reduction on any significant scale would ultimately reduce production, if traditional farming methods were retained. One could expect therefore that if India maintained its present agricultural system, but somehow halved its population, then its farm productivity would fall correspondingly or near to it. It is the traditional farming method that is at fault, not too many people.

The population of Ireland was about nine million prior to the Great Famine of the mid-19th century. Population declined through premature deaths and migration to about two million. Today Ireland's population is about three million. This has not resulted in any great burgeoning of agriculture. On the other hand, the Netherlands is in the same climatic belt with not too dissimilar soils. The Netherlands, with a land area of approximately 15,000 square miles, is less than half the size of Ireland and possesses slightly more than 13 million people. With some 866 people per square mile, it is among the most densely populated nations on earth. Yet it has a most efficient agriculture.

Even the most ardent Irish nationalist would have to admit that his agriculture lags in comparison. Holland's hog raising is among the best in the world. Dairy cow yields are among the world's highest; the same can be said for horticultural yields generally. Despite this high base, Netherland's agriculture continues to grow at the rate of 2.2% per annum. These imbalances between contiguous countries which are not tied to population densities can be traced throughout many regions of the planet.

The real issue as regards population and agriculture is that the planet's current population is not great enough to pressure agricultural *potential* at any stressful level. Projections indicate that world population will double by the beginning of the 21st century. If we want, we can double agricultural production in the next ten years and go on doubling it every ten years, until we have at least a six-fold increase of our present production; and that figure is probably too pessimistic. We shall return again and again to the issue of population and agricultural potentials. For now, however, we must continue with our analysis of the inter-relation of population and resources, and the human dynamic of those who comment on it.

This brings us to a second reason for the truly incredible errors of the crisis brokers. Their data always seem to exclude humans as active participants in the process of living. Human adaptability, ingenuity, flexibility, and human propensity for change — often rapid change of viewpoint and activity — is simply discounted. Yet the whole history of man shows that it is those qualities that move human affairs, give them their direction and dimension. Humans are statistics only in a very general sense and they are always refuting statistical projections, especially over historical time.

Once when I showed a farming friend — a wise, practical and competent man — some ten years ago, a particularly gloomy but typical article on agricultural disasters impending in the U.S.A. (but which did not eventuate), my friend made an apt remark. "Obviously," he said, "slobs like me don't count!" Yet to me my friend represented the human dynamic of agriculture. Incidentally, this article which predicted the end of the family farm in the U.S.A. was not only wrong conceptually but has proven wrong in the event.

The thesis was that the owner-farmer was disappearing as large agribusinesses devoured privately held farms. Again a linear projection was at fault. Many privately-held farms remain financially viable and are unlikely to vanish if not artificially submerged. There is a good reason for their viability. In those times of penury, which afflict farming all too frequently, the individual farmer has the capacity to hang on and endure. He will accept a 1 or 2% return, even a loss, yet not give in because the farm is his livelihood, home, and love. A large corporation, especially if it has public shareholders, cannot take its tiny profit margins, or even losses from time to time, without encountering severe investor backlash. Compared to the individual farmer, publicly-held farming corporations are fragile growths and as such should not be regarded as too fearsome. It might be further noted that the family farm has proven its viability because of the human factor, the disregarded factor.

The family farm does however have two formidable enemies in the U.S. They are co-joined. Inflation is pushing farm values to artificial levels. Estate taxes based upon these high levels, in practical terms, mean that about every 30 to 40 years each farm must provide the Federal government with substantial sums in liquid cash. No family farm can produce such liquidities from earnings. Instead they go under the auctioneer's hammer.[4]

In summary, the best that can be done in projecting humans as statistics is to sound warnings. Let us accept these warnings, but accentuate the ability of humans to react to crisis creatively and ingeniously, if they are allowed.

We must probe deeper as regards the contemporary crisis broker, for we have not finished with him, nor he with us. Intellectuals in the West, and particularly in the U.S., are seeking power and influence in government. Having defined the problems to their satisfaction, they want to be part of the remedy. Here lies a new outlet for their ambitions. It seems they are driven to save us from ourselves and then manage us so that we do not commit error again. By the formality of their education and by the narrowness of a specialty (more intellectuals than ever are specialists), many intellectuals cut themselves off from life as it is. They have great difficulty in seeing how things work. The greater bulk of their lives is spent within the academy as students and teachers. The academy is not representative at all of the outside world, least of all of our point of interest, day-to-day agriculture, particularly agriculture in its planetary dimension. In the detached academic environment anything becomes possible, everything is practicable and without hindrances.

Often intellectuals are passionate men and drawn to what might be described as a form of "petite millenarianism." They see mankind confronted by a kind of material damnation. They seek and then recognize a "Devil Image." For this kind of intellectual, the "petite millenarianist," the devil appears in tangible form. Capitalism, the Corporation, overpopulation, human greed, imperialism, multi-national business, supersonic jet airplanes, urban sprawl, feeding corn to cattle, highways, super tankers, pipelines, refineries, disposable containers, plastics, even the personal automobile, all become tangible devils. Devils have to be exorcised. But this can only emanate from those who know, those who have found the way. He that has the way can assume the role of prophet.

Fortunately for us, as opposed to prophets of more robust millenarian movements that have plagued history, our intellectuals are normally gentle souls. They confine themselves to preaching how we ought to save ourselves. They normally restrict themselves to words. They are verbal Messiahs rather than active Messiahs, but the messianic notion remains. "I have seen the world moving toward perdition. I have found the way to save it. Listen, O people." There it used to end.

True millenarian movements arise when people are sorely afflicted by real and tangible crisis. They occur when that crisis is so great that the very life way of the community is split asunder, when cultural underpinnings are ripped up. Lenin, to describe this state, created the word *deracination* from the French word, *deraciner*, to "pull up by the roots." Our "petite millenarians" however, are not enduring any such trauma. On the contrary, they are born of an affluence, born in turn of a success in most forms of industrial and agricultural production. In the West and especially in the U.S., the great bulk of us (and especially our intellectuals) have never been more affluent. In a perceptive piece published in *The London Economist,* Norman Macrae offered two reasons for England's "drift from economic dynamism" which began about 1890.[5] "First, the upper classes began to regard business as something vulgar, and to look upon new factories as things that were ecologically unfair to their pheasants and wild ducks. That," observes Mr. Macrae, " is exactly the mood of America's intellectual upper class now." His second reason for the decline was as follows:

> **After about 1870 a progressive person in British public life no longer meant a person who believed in progress, or a person who was eager to go down to the roots of the ways of doing things in order to cut and graft wherever an improvement in production or effectiveness or competitiveness or individual liberty could be obtained. A progressive person began to mean a chap who did not like progress and change very much.**

Macrae again states that this is exactly what is happening in America "in these years just prior to 1976." He continues that, "if this gloomy view is right, world leadership is likely to pass into new hands early in the next century."

Without doubt, if we do heed the particular intellectual group under discussion, Macrae will prove to be dramatically correct. Fortunately, intellectuals are rarely actors and therefore real results do not always emanate from their words. This seems to be especially true in the U.S. They have not yet had a material effect on agriculture, which in most of the West and in some parts of Asia, continues to burgeon.

It would seem that their words have produced a psychic impact upon segments of the people of Western nations. One thinks of the uninformed despair of many of the young, with whom intellectuals have direct and increasing contact. Adults should try to put themselves in place of the young. As adolescents, eighteen years of age, when life should present its most exciting and stimulating challenges, they are told that all is hopeless. They are informed that there is little they can do except strike out, punish, destroy whatever "devil image" a particular intellectual holds responsible.

Now, some institutions *are* failing. Paradoxically the failing institutions are not those condemned by intellectuals, but those which intellectuals would like to enhance, those central elements of government which restrict, rather than free. To return to our young, we must pity them as they are denied the optimism of youth and their right to form visions of new futures. Let us try to free our children, as we try to find out why agriculture is also being fettered.

The second psychic impact arising from the cacophony of impending doom is a more general loss of confidence. Such a loss does not directly affect agriculture — not yet. By nature a farmer relies on himself and his relationship to plants, animals, soil and climate. He recognizes that he is the prime mover in the symbiosis. But a farmer also lives in a general community, a community of politics, government, bureaucracy, and his fellow citizens who make up his culture. He must relate to bankers, manufacturers, transporters, processors and consumers in the general marketplace. It is here where lack of confidence intrudes, creating misgivings about the future. No nation, no culture which seriously doubts its future, can seriously expect to have one. Thus, while our would-be saviors, in the intellectual world may be having little direct effect on the production of food and fiber, they may be having a significant indirect effect which may come to be translated into a direct effect. But confidence can be restored. Youth can be rescued from uninformed despair and have returned to them the privilege of shaping the future.

In this book we have embarked upon an exploration which will lead us to doing just that — as far as the basic ingredient of agriculture is concerned. As a first step let us examine some of the premises upon which the prophecies of disaster are based.

We have noted the contemporary phenomenon of basing prophecy on linear extrapolations in terms of population. Gifford Pinchot, who in many ways could be called a parent of the conservation movement, wrote in his pioneer book, *The Fight for Conservation,* that as of 1910:[6]

> **We have timber for less than 30 years...anthracite coal for but fifty years....Supplies of iron ore, mineral oil and natural gas are being rapidly depleted....**

Yet more than 10 billion cubic feet of timber was cut in 1965; the 1973 cut was 6% higher, and all indications are that timber production is still increasing at 1 to 2% per year. With continuing successes of tree farming, there is no reason why this production rate cannot be maintained at increasing levels, and this raises the question as to why Pinchot could have been so wrong.

The answer would seem simple. Like our contemporary prophets relating linear extrapolation to prophecy, he totally neglected the human factor. In doing so he introduced an arbitrary and incorrect rule. By denying human input, he said in effect, "There will be no replacements and no substitution in my equation." Of course there have been both. While forestry may not be our most advanced agricultural pursuit, remarkable strides have been made in these past fifty years in general tree culture. In fact, in the U.S. we now plant more trees than we cut. In addition to this we have the factor of substitution. Many articles previously made of wood are now constructed of other, usually cheaper, materials. Logs are "peeled"; the resulting "peelings" (and a tree produces thousands of such peelings) are glued to lesser materials to create beautiful panelings, actually offering a walnut veneer (or whatever type decided by use of differing parent trees) at a fraction of the cost of a walnut board. This, incidentally, is also a form of conservation.

The human dimension which has been omitted can be classified. The primary issue is change induced by science and technology, which in turn induces more change, and still more, in a given direction. The direction results from making good on a need or replacing a scarcity. This changes, too, inducing yet more changes — including more scientific and technological change. All of this dynamic interaction arises from human beings operating under quite definitive conditions. Primarily necessary to release this human interacting are: knowing the challenge; having freedom to respond to the challenge, and having an incentive to make it worth the individual's effort to respond to the challenge. Omission of this human dynamic from any forecast makes that forecast irrelevant. In forecasting one must say, "Here is what might happen. If this does happen, what then will be the human response?" The last sentence is always omitted by the doom-sayer perhaps because such forecasters have little knowledge as to how the world of action works.

Let us also observe a cautionary note. The widespread American belief that science and technology will take care of anything and everything is almost as dangerous as its opposite: that science and technology will bring us to disaster. One can agree with Arthur Clarke that, "Any sufficiently advanced technology is indistinguishable from magic,"[7] and yet realize that magic also has limits. There are ultimates that even technology cannot touch, especially in terms that are economically viable. We need to develop a new pragmatism which falls between the dire predictions of the doom-sayers and the unquestioning optimism of the "can do" technologist. Here is the real challenge for the intellectuals. Let them explore this new middle way as compensation to us for their previous negativism.

Another factor in linear projections is their prophecy of the end of a particular resource. We cannot ignore this. It is one thing to project the end of a renewable resource such as lumber and be wrong. It is yet another thing to predict the end of a resource that is not renewable and which exists in a finite quantity, such as, say, oil or iron ore.

There is no doubt that one day the planet will run out of oil and, possibly, iron ore. If the Second Law of Thermodynamics is to be accepted, ultimately the planet itself will become a dead hulk, a lonely, deserted orb spinning in space. In any geological time span, man is up against it. The universe will neither care for him nor ensure his survival, nor even give a shrug at man's passing. He is brother to millions of species that have gone on to extinction. This is why man is so important.

If the Second Law is true, no one can stay it. Man's only alternative is, long before his extinction, to escape to another planet or to make a giant leap of faith such as is asked by Teilhard de Chardin. Teilhard, a scientist as well as a Jesuit priest, sees man evolving until he merges with the Godhead. Man is then no more. Although Teilhard never liked it, the same idea is central to doctrinal Hinduism and, in a refined form, is central to Buddhism.

But the key feature in all of this is man, not the material environment he finds himself in at this moment. The dinosaurs and all those other species which have disappeared could never escape from their decaying environment. They simply lacked the capacity. With man we make an enormous, indeed, unfathomable and unbelievable quantum jump in capacity. It all comes from his mind: fusion power, fission power, electrical power, mechanical power and, transcending them all, power to create and recreate. This is why man must be central to any projection made about his future, immediate environment included. Statistics about his environment are fine and make handy reference points — sometimes even sound warnings. But that is all. Therefore let us return to oil and iron and their imminent exhaustion.

Oil as energy displaced the horse, mule and ox on the American farm. If, in turn, the horse, mule and ox are to displace oil as energy on current scales of use, just to grow feed for these animals would require about a forty times increase in our croplands if we attempted to maintain current scales of agricultural production. This would be impossible. Thus animal power, and production, would fall dramatically. Yet linear projections show the ultimate unavailability of oil in the quantities now used for agriculture. Neither does it matter if these projections are wrong in terms of nominating the final day. It really is not of too much consequence whether oil runs out in 2030, 2060 or 2099. Something has to be done. We can despair, or we can find alternatives: new energy, new techniques.

As far as iron ore is concerned, the situation seems easier. Only recently has U.S. begun large-scale metal recycling; the impact of this has yet to be accurately assessed. But steel and cast iron are important on the farm, in agricultural processing plants and in agricultural transportation. Should these become so scarce that their cost becomes economically unviable, the task will be to find alternatives.

The suggestion that the remedy to shortage lies in finding alternatives will not easily satisfy all people, least of all our irascible crew. To date, their arguments look impressive. They have produced masses of statistics which all reach the same conclusion, that demand is outrunning supply at rapidly increasing rates. How does one respond to this beyond the flat statement that we must find new equivalents? Let us see if what happened in the past is any guide to the future.

We are reluctant to enter into the future-predicting game. No one in previous times could possibly have predicted what is happening today. It just could be, therefore, that no one today can predict the future. Who in the year 1885 could have given even a faint facsimile of what American agriculture would be like in 1985? In 1937 the predictors said the U.S. had 41% of the world's oil reserves and the Middle East had 30.3%. By 1960 the ratios had changed to 61% for the Middle East and 10.4% for the U.S. New discovery changed the picture dramatically; no doubt such changes will occur again.

In 1950, the year in which Pinchot said oil in the U.S. would be depleted, production was 1,944 million barrels with a proven reserve of 25,268 million barrels. In 1970, production had increased to 2,471 millions barrels and proven reserves to 31,613 million barrels. Let us assume that these disparities between prediction and actuality are but temporary aberrations — let us assume our oil will run out.

History indicates no particular evidence to suggest that the dynamic between man and his resources has or will disappear. To the contrary, man's ability to innovate and create as regards resources is improving, not diminishing. This improvement comes about from evolution, which as a force seems to accelerate, and from technological change born of accumulating knowledge. We do not know what the advances will be these next one hundred years, and here we are at a disadvantage relative to our irascible friends, in terms of "proving" our point.

Graphs look impressive as they show this or that change projected into this or that distant future. The graphs showing population expansion at rapidly increasing rates in conjunction with other graphs showing resources diminishing at rapidly increasing rates, can be as discouraging as they are graphic. All we can respond with is the human dynamic, difficult to quantify, neglected in most projections, and proven only by its record in the past. But what a record this has been.

Not only are new mineral deposits being found, but new processing methods make the working of neglected deposits economically viable. Geological knowledge increases and so does our ability to find and use new minerals. We have not really begun to search seriously for new energy sources in the U.S. primarily because the human dynamic that has served us in the past is being inhibited mainly by government.

The list of man's evolving capabilities is long, perhaps endless, and cannot possibly even be summarized here. Even those classical economic rules of "diminishing returns" can be questioned, if not overturned. This classical view was that increasing improvements in production systems must gradually but inexorably show an increasingly diminishing return. Is this any longer true? The advent of new industries in knowledge; technology; systems as a productive device; more effective use of investment capital and operating finance; all suggest that no longer need returns diminish as production method improves — if this in fact were ever true. It is in agriculture where the most striking changes have taken place and there is no evidence here that this rate of change will stop or even slow down; more likely, it will be enhanced.

Let us look at some improvements and sound a note of caution against becoming too optimistic. Either side in this argument should avoid ultimates in logic, because in agriculture there are few ultimates and these are easily discernible. Agriculture is a dynamic that is constantly changing, fortunately toward greater production. But there are limits, mainly concerning animals rather than plants. Animals probably have quite definite limits to natural increase and physical growth, and these are discernible. We are probably coming close to them now. Plants also have limits, but in many plants these are much further away. Let us look at an example in the animal world.

In the year 1800 most of New Zealand was covered in forest. By midcentury substantial proportions had been cleared away, often by destructive and unnecessary burning, something New Zealand rues to this day. In the end the New Zealand farmer made good for his rapacity by such means as the development and nurture of animals. On a typical New Zealand bottomlands farm (which had been cleared by burning) fertility was initially very high. This fertility, based upon the residues of burn, quickly disappeared, just as it does today where primitive agriculturists practice "slash and burn" agriculture. With fertility reduced, one acre in this New Zealand farm carried about one ewe. This sheep yielded about two to three pounds of wool per fleece and about 30 pounds of meat. Mortality was fairly high from diseases and other causes. Then came a steady development in sheep genetics, giving rise to massive increases in production of wool and meat, and in disease resistance. Most significantly these enormous yield increases were produced in less than half the time hitherto, normally a little over one year of the animal's life.

Of equal importance, in New Zealand there was a corollary development in grass genetics, grass husbandry, soil drainage and fertilizers. Today, a century later, the same bottomland carries nine ewes per acre. Fleece yield is nine to ten pounds. Yield of meat has doubled or tripled. In direct contrast, yield of meat from a slaughtered sheep carcass in the Sahel region of Africa is less than ten pounds. No figures at all are given for wool because its impact as a fiber is negligible. To reach this status, the animal requires four to five years of life. In those few places where agriculture is pursued seriously, such as New Zealand, such incremental increases over the past century are not unusual. There are very few places where agriculture is pursued seriously.

We would be gravely in error if we now went on to make the same mistake as do those who extend their data in linear extrapolations. The carrying capacity of those New Zealand acres which increased more than nine times in a century is less likely to increase nine times again in the next one hundred years. Such will certainly not occur if based on grass. Perhaps, if New Zealand decides to utilize its excellent sorghum-growing capability, such an increment might be realized. It is in nutrition that the greatest potential for gain exists. But the animal itself is probably reaching its natural potential limits. Its yield of meat could double again, but such is unlikely. It is most unlikely that wool yield would make the same quantum jump these next one hundred years as it did over the last century.

Thus, further augmentation of sheep on the same scale as has occurred in the past is unlikely while the animal retains its present conformation. While a sheep retains its "sheepness," to put it in Zen terminology, it will increasingly resist further dramatic change. A new animal would have to be developed which might be related to a sheep, but yet would not be a sheep. This same situation exists with all other animals. How far their yields can be developed is limited, but we have not stopped creating. In the U.S., for example, the African eland might be developed for use in the American Southwest, but the economics of the proposition have not been definitively explored.

To return once more to New Zealand, here we do find a new food animal with important developmental prospects. During the last century British settlers introduced the Scottish red deer to New Zealand. These animals began to run wild, multiplied into great herds and became a serious menace to the native forests. In many a beautiful glade in one of those magnificent New Zealand forests, a mantle of young trees covered the forest floor, part of the natural regeneration process. However close examination would usually reveal that regeneration was an illusion. Deer, almost with surgical precision, would have removed the topmost shoot of every tiny sapling, dooming each to a stunted, living death, bringing later death to the grander forest.

By the turn of the century, the New Zealand government was paying bounty hunters to destroy thousands of deer. In wild and rugged terrain animals were left where they fell, for nature to reclaim its own. Sometimes, animals were skinned and the hides eventually transported by pack horse.

After World War II came the helicopter. It became possible to retrieve the whole animal and a thriving venison market evolved. It was but a short step under market stimulus to look into further dimensions of red deer. It was found that not only did they utilize grasses with about 50% more efficiency than sheep in producing a pound of meat, but that they could accomplish this on a wider variety of grasses. The carcass also proved to be more efficient in that it held much less fat and therefore less waste than either sheep or cattle. A deer farming industry has now begun. If history is any guide, even greater efficiencies will be induced utilizing red deer. At least two countries, Taiwan and Ireland, have now imported breeding animals with a view to farming them. Thus, even with animals, the dynamic continues. Red deer will doubtless reach their potential within the agricultural symbiosis.

We really have little idea how much further we can advance plant development. We do know that with plants, potential is very much greater. In the mid-1950's an Indian researcher stated that sugar cane yields in India could theoretically be increased by 1000%. While sugar yield has increased dramatically since then, it has reached only a small fraction of its predicted potential. Much the same can be said for most plants now in use by man, although few could be anticipated to have a 1000% potential for increase. Further potential for enhancement of plant use rests in developing new varieties suitable for human and animal consumption. The planet possesses nearly ten million plant species, but less than 100 of these are in common use as human and animal foodstuffs. The potentiality of this enormous ten million species "gene bank" is largely unexploited.

Soils, too, are limited, not only in quantity, but because of mineral contents that prohibit agriculture. As far as water is concerned, irrigation can counteract climate to the degree that water is available. Temperature, which is so critical to many plants, can hardly be controlled except in quite expensive enclosures.

The significant factor, however, is that these limits have not yet been met anywhere — indeed, we really do not know what the limits are. The other more important fact is that high yields occur in only small pockets of the planet. In other words, rather than coming up against the issue of limits to plant and animal growth, we face exactly the opposite. We face a largely untapped resource.[8]

The same imbalances that exist in animal production in various parts of the world exist in plant life. We have noted some of these, especially disparity of rice yields in Asia. If these imbalances were to disappear, the planet would not merely relieve its current food shortages, it would face enormous surpluses. China, for example, grows more than 31% of the world's rice, the largest producer by far of this grain. India produces 21% while the U.S., now the world's largest exporter of rice, produces less than 2%. If mainland China were to increase its already reasonable rice yields to those of its neighbor Japan, an enormous glut would ensue. China currently produces approximately 66 million metric tons of rice in a good year. This yield is adequate to meet the needs of the rice-eating segment of China's population. If Japanese yields were to be equalled, China would produce about another 30 million metric tons. Most of this would be surplus to China's human needs. Against this potential surplus of 30 million tons, the U.S., as the largest exporter of rice, exports less than three million tons. Thailand, the next largest, exports some two million tons per annum. Exports from other countries are small. The propositions based upon linear extrapolations however, tend to ignore these data on production, which usually, in their lexicon, remains static, or even decreases, while general consumption figures increase. The resulting imbalances look frightening — and would be if they were truly based upon reality of human conduct and took even minimal notice of human potentiality.

But we have not yet come close to limits of agricultural growth. We do not even know what these limits are, except that they are higher than what we know we can do now. This does not guarantee these upper limits will be reached. Indeed, it probably will not be done. We shall as a planet fail to develop our agricultural potential. Reason for this failure will not be too many people related to few resources or that developed countries are wasteful. It is developed countries which are producing not only surpluses, but also pioneering research and development for new advances. The reasons for our current failings and potential failings are quite different. If we fail to analyze a problem properly, a solution to that problem can only be found fortuitously — and that is a poor way to resolve anything.

To date agricultural results have not been particularly hampered by the pollution argument, except for its bearing on pesticide use where a serious situation is arising. But the portents are about us. Those interested in food and fiber production and its increase should realize that soon efficiency may be affected, especiallly food production in an advanced state of agricultural technique. The inefficient, in this new world of restriction, will never get off the ground.

Yet another principle might be observed regarding pollution and agriculture. Pollution is really inimical to a well-run farm. Pollution here can be defined as disposal of wastes indiscriminately, deteriorating the natural habitat of plants, animals, soils, climates and, therefore, man. In a recent discussion on this issue of curbing waste in agriculture an associate observed, "On a well-run farm does one not curb waste — there is *no* waste."[9] In other words, a well-run farm does not pollute.

This was not always the case. Archeological sites of ancient peoples are often identifiable by their middens. Each home had one, a place where wastes were thrown, as were worn out garments. Even the skeletons of infants are sometimes found in these refuse heaps, reflecting a high infant mortality rate. In contemporary times, primitive societies cling to the practice of indiscriminate disposal.

The same practice among tribes in Southeast Asia no doubt goes on today. Hogs, dogs, and chickens scavenge middens, but the accumulations grow and the people concerned have little consideration for the impact of their wastes on anything. Man is a long-time polluter. He has never concerned himself about it. Most humans have not changed much in this respect.

In agriculture, but only in advanced agriculture, there has been change. It is now simply too expensive to waste. Improved management and control techniques inhibit waste. But this is only part of the contemporary story. Something else has changed. Today mankind pollutes more and more dangerously, but not in terms of waste. Rather, it is by his techniques to enhance production. There is no question but that today certain of these kinds of pollution must be limited. The question is one of degree. Thanks to environmentalists, man is being made conscious of his bad habits and for that reason we can award environmentalists an accolade. But being aware, and taking appropriate, prudent action commensurate with reality, are two entirely different things. Here we cannot be quite so laudatory of environmentalists.

The issue has now been turned into an adversary relationship, a "we" and "they" situation. "We" are against pollution and are certain of our grounds, so certain that we prescribe remedies that are didactic, arbitrary, and uncompromising. "We" are not interested in the interrelationships of our proposed remedies with other elements of human existence, only in attacking pollution in a particular way. Not only must such "pollution" be stopped here and now, but past effects must be "cleaned up." "They" are assigned the role of villains, rarely identified by name, just as "polluters" and that is enough. The fact that, just as with primitive man, everybody pollutes, is usually glossed over or rationalized as being part of what "they" have done to us with their technology. As could be expected, politicians did not take long to enter the pollution arena. A new factor thus enters, not that of necessarily being for or against pollution, but rather that of political self-interest. This in turn introduces a haphazard or random effect into the equation of "we" and "they."

A general illustration of the dilemma and probable results of the adversary relationship can be provided by using pesticides and other chemicals, factors of extreme importance to agriculture. Pesticides are critical to farming to curb predatory rodents, insects, weeds, and diseases. In the U.S. it is probable that without pesticides, agricultural productivity would decline 40% or more. In Kansas, the bindweed if not countered, would reduce sorghum crops by 78%; oats by 36%; barley by 39%, and wheat by 30%. Bindweed is but one inhibiting perennial in Kansas. We could travel throughout the U.S. and find similar examples.

Without pesticides, agricultural surpluses of American agriculture, so significant to U.S. economy and important to the world, would disappear over one growing season. In countries where agriculture is marginal, pesticides offer the first mechanical step in increasing usable foodstuffs. Pesticides also curtail diseases which affect farmers' productivity. Elimination of malaria through DDT is a notable example.

Other chemicals are related to pesticides. Their use in agriculture has yet to be fully exploited. Chemicals can aid the process of photosynthesis in plants. Experiments with soybean, that old, old wonder crop, are now under way. At present soybean leaves use only about 3% of sun's energy which is converted into food calories. It seems highly probable that using appropriate chemicals, this collection rate can be increased to 6%, resulting in double soybean yields.

Soybeans are but one crop where this process would apply. In fact, any crop like soybeans, which is an average converter of sun energy, can yield equally spectacular increases. There are other chemically induced possibilities. These range from persuading non-nitrogen producing crops to produce nitrogen to add to soil as do clover grasses, peas, beans and soybeans. Chemicals can help plants make better use of available nutrition. Plants tend to be quite wasteful. It is estimated that they use only 50% of nitrogenous materials, the rest being discarded.

Yet another field is use of chemicals to control ripening. The avocado, for example, naturally spreads its ripening process over a considerable period of time. This has an appreciable effect on agricultural operations as far as avocados are concerned. Farmers can extend harvesting season and obtain a more economical use of their labor. They can spread market risks. Waste can be cut to a minimum. Most fruits do not ripen this way. Ripeness comes with a rush and it demands immediate picking and marketing. Such rapid ripening gives rise to an enormous potential for waste and often gluts the market. Work to regulate ripening by chemical means is well developed and offers benefit of another agricultural advance, a spreading of the "avocado syndrome." Controlled ripening is really a process of growth regulation. A further extension of the process would be to curtail plant growth, control the ripening process and possibly thereby reduce its consumption of food per unit yield.

Use of pesticides and other chemicals is too important to be put into a "we" and "they" context of adversary relationships. All effective pesticides are highly toxic and therefore dangerous. Their effects relative to their primary mission, and their side effects, need not only to be analyzed *a priori*, but monitored during operation. Such a process is not beyond us. It happens every day. Cooperation between parties is the way to get the benefits.

If the present adversary relationships continue, it is obvious what the results will be. We cannot exist without pesticides so they will continue to be produced. Pesticides are almost always compounds. Therefore a manufacturer could, knowing full well that he is a "they," anticipate challenge by the "we". Assuming that any new compound will have a five year operational life before it is banned, he could simply stay one step ahead of the law by continuously altering compounds over five-year cycles. Such a system would be both irresponsible and dangerous. Should it arise the cause will be not only the rapaciousness of the chemical companies.[10] Rather it will arise because there will be no viable alternatives to agricultural pesticides. They must be produced and used. The important fact is, however, that the dangers inherent in all pesticides can be nullified by proper use. These dangers have always been over-dramatized and exaggerated. In short, we can live well, if not better, with pesticides than we can with all those other risks and dangers inherent in being alive, provided that we behave sensibly. The whole controversy is symptomatic of where zealotry of the irascible ones is leading. We do not have to follow. We can all act more sensibly than this.

The same situation applies to pollution in general. Are penalties the way to resolve these issues? Must everyone be punished — the producer, the user and the consumer? This is essentially a primitive religious view where sin is never atoned for, but only earns punishment or damnation on a permanent basis.

In agriculture, other potential penalties come less directly. In the end these could become serious inhibitors, but the task now is to sound a warning rather than outline a reality. In its report of 1977 the Council on Environmental Quality stated that capital expenditure by the U.S. *Federal* government for air pollution control legislation ought to be $153.1 billion from 1976 to 1985. The cost by 1985 was estimated to be $20.6 billion per year. This is only the beginning. This only concerns air pollution.

The National Commission on Water Quality estimated that to meet waste standards already set by law, some $116.8 billion will be required from 1976 to 1985. Over all, this direct government-sponsored capital investment will amount to $269.9 billion for the years 1976-85.

There are additional charges which cannot be estimated. Every business or agricultural operation that becomes involved in the actual process will also carry heavy administrative costs. Pollution control attempts could exceed future education or defense expenditures.

No one knows where the greater burden will fall. Farming today is at best a marginal enterprise. Any overloading of agriculture will result in further disruption of the fabric of agricultural operations, higher prices, or both. This raises the question, not only for farmers, but for everyone, what will be the benefits?

The National Council on Water Quality assessed material benefits of clean water at about $55 per household annually. This compares with water cleanup cost of about $500 per household per year. Not even wealthy Americans can live with such inverse ratios, especially when we realize that the direct cost of air cleanup, in like inverse ratio, "is following along behind." Then we must add indirect or hidden costs which individual operations will have to bear. Beyond that, there is the escalatory tendency of any legislation of this magnitude. Costs escalate because of wrong estimations, and from "add-ons", because government operations grow and rarely diminish. Yet we still have not answered our question, what are the benefits.

The findings of a perceptive piece[11] which almost exactly parallel what has been said here, estimated, that in one factory the cost benefit of removing sulphur dioxide emissions was 1.7 cents per pound of product. The cost of removal, however, averaged 22 cents per pound. Extrapolating on this single issue, the author assumed a national cost benefit ratio of 13 to 1. Then an amendment to the Clean Air Act, illustrating the escalatory nature of government operations, was estimated to push the cost of emission control up to 57 cents per pound of sulphur dioxide. The cost of removal, however, averaged 22 cents per pound. Extrapolating on this single issue, the author assumed a national cost benefit cannot bear them at all.

Thus we can say that the environmental movement is promising to change U.S. economic balance greatly, and thereby, the economics of producing food, fiber and shelter. There is no guarantee that under the measures envisaged, air will be much cleaner than it is now. Waste might fare better, but not much better.

Should this prediction prove correct, that benefits will be all but indiscernible, will the zealots redouble their efforts? This is how failure normally impacts on zealotry. Perhaps this is the best way to go in the long run — to prove by bitter example that "pure" air and "pure" water are not worth the cost even to the most dedicated of us. But the cost, in any event, will be high. The entire economy will be hurt, agriculture along with it. Capital, which might have gone into irrigation, plant and animal genetics, agricultural chemical and fertilizer research, rural credits for new land development, will instead be dissipated to attain a small result which it is far from certain most of us want or need. Environmental enhancement can also become a punitive measure. This illustrates the mind-set of the irascible ones. We might, therefore, seek ways to divert this well-meaning drive into more constructive channels where cooperation, rather than punishment, is the *modus operandi.* Before that can be done the environmental movement will have to come to terms with technology; not just agricultural technology, but technology as a whole. Here is a challenge!

Few people who have read Jacque Ellul[12] could fail to be impressed by either his arguments or the gloomy brooding that overhangs his long and often convoluted work. In a sense Ellul synthesized the work of anti-technologists from the Luddites to the present time. He also gave to technology (*"la technique"*) a personality compounded of enslaving efficiency and dominating omnipresence. One would have to recall Stanley Kubrick's "2001: A Space Odyssey" to experience a like situation, as in the scene where the computer HAL with his assumed human personality gradually, but inexorably, takes control of not only the spacecraft but its human occupants.

Like *la technique*, it is HAL's efficiency, his pandering to human comforts and foibles, his indulging but always concealed and absolute power, that in the end promises to make him master. Kubrick relieved this tension. In the end HAL was dismantled by a human. With Ellul there is no dismantling. Humans are not actors...merely victims. *La technique* is inexorable in its drive for mastery over humans, binding them to it because it offers them so much. Then they become enslaved because they cannot live without *la technique.* The baubles and trinkets it produces become the beginning and end of human existence, determine not only the quality of human existence, but of human nature itself. "It is already too late," says Ellul, "it has already happened. Technology is now a force on its own. It is the master; we the slaves. We are Faust in modern form. But Faust does not die. It is much worse than that: we, as Faust, remain enslaved in perpetuity."

Humans love such a tale. For adults it offers the vicarious thrill that the more ghastly aspects of Grimm have for a child. But everything does not come out right in the end. Indeed, as at least a dozen American writers have suggested, following after Ellul, it all gets worse and worse. Being American, however, the tale can never end in black tragedy. There has to be a way out. The way out is to give up technology, to return to a simple life of "harmony with nature," a sort of Rousseauean dream world where nature becomes the mother caring for her children, more so now, because though they erred, they have returned to the fold. It is the Garden of Eden syndrome.

There is no doubt that an atavistic yearning for life in Arcady lies within all of us. There has been too much of it among us, this yearning for a perfect past, to have many doubts. This modern atavism is no different from its predecessors. It is born of the strong passions of a fear and of a hatred, in this case, for technology. Technology is the "Devil

Image" which must be exorcised if mankind is to be saved. There are remedies, American remedies as expressed by these anti-technology writers so we may "return" to nature. One, a Mr. Charles Reich has even offered as his solution that America should "liberate" itself by raising its consciousness to that demonstrated by the youth movement of the 1960's.

These writers all talk nonsense. In common, they know little of technology or work, of real interrelationships between people and technology, and absolutely nothing of the courage of man. What we are witnessing among them is a simple atavistic movement. These movements are ever common. They are well documented for Europe in the 11th through 16th centuries. they are accurately described in many writings covering other parts of the world.[13] This late 20th century movement is interestingly different in that it is inspired by persons who have spent their lives mostly in academia. Most atavistic and millenarian movements in the past have involved the opposite, at least as far as followers are concerned, i.e., peasants and landless laborers. But the ingredients are basically the same. There is a tremendous feeling of insecurity and an absence of roots as a centerpoint in one's life. There was a certainty that a definable evil was abroad, giving rise to one's trauma, and a sense that they, the insecure ones, must rise against it and strike it down for the good of all mankind.

Already in this chapter we have spoken of this remarkable, recurring phenomenon. This millenarianism and its cohort atavism run like isomorphs throughout human history. But the uneducated millenarians or atavists were somewhat different from our educated ones. They looked forward, or backward, to a golden state where the problem of poverty in particular was solved for them in perpetuity. Some degree of affluence was for these poor folk a remedy for their ills. Our educated ones are already affluent. In the main they imply they do not like affluence. They often claim that they have been betrayed by technology itself into a world where man is "de-natured" by the affluence born of technology. Their "awareness" has been stultified by urbanization, highways, mass sports, mass entertainment and especially TV (as though one had to watch it, consuming things one does not want). They are continuingly manipulated by an artificial environment created by The Establishment and by The Establishment's lackeys, the technocrats.

Presumably when we return to a more "natural" existence, where all these appurtenances of technology have been abolished, the great Mother will care for her redeemed children. The primitive tribe and peasant societies seem to offer a great attraction to these atavists. They romanticize the "rhythms of the earth" and the harmony of a life close to nature which is the fortunate lot of peasants and primitive folk. The peasant is not dominated by external forces either, but can shape his own work day according to his own determination. The peasant lives in an equilibrium, bucolic, carefree, unburdened, unwanting but receiving, so his modern admirers seem to be saying.

Here one must stop this analysis of these yearnings. It is too much. These educated ones know nothing of primitive peoples and peasants, neither do they seem to have understood the history they have read. They are living a mythology. But this detestation of technology does not go away. Thousands upon thousands of people have been affected by the story as it is told in fine prose, with intellectual underpinnings and a certain aesthetic finesse.

Technology is part of man — it is part of peasantry, even part of the life way of primitive peoples. The first man to knap a flint tool was a technologist equally as skilled as a lathe turner involved in precisions of the aerospace industry. Technology is not only work of man; it is in the nature of man. The normal demonologies are aggression, passion, greed, hatred or unforgivingness. Technology, on the other hand, is passionless in terms of its own dynamic, causing harm only when man in his aggressions, passions, greed, hatred and unforgivingness decides that technology will be used to further these ends. To describe technology as a force in itself, without man as the mover, is to belittle man's power, good or bad, and create a demonology which in fact does not exist.

Technology is blamed for much else. It forces people, it is said, to dull, repetitive and even exhausting work. The implication is that Nature does not impel people this way. Again, what nonsense! Nature is the harshest of taskmasters. Weeding a field of turnips is repetitive, dull, exhausting, especially if one is endeavoring to race the weather (because Nature can play the game that way, too, and not obey the rules). More importantly, we now have technologies that do not require dull, repetitive, or exhausting work by man. We have technologies that do not need man at all. We do not hold such a mastery over Nature. The biological cycles of animals and plants make demands with a disregard for the likes or dislikes of man. Soils can be stubborn and weather capricious. Heat, cold, wetness or dryness of fields may seem to be a harmony to the city dweller, but such is indiscernible to him who is in the field. When they talk thus of Nature, our anti-technologists are dealing with romantic unreality.

The anti-technologist also condemns contemporary technology for pushing people into buying what they do not need. It is not people, they say, who are greedy to possess and unwise in their choices; rather, it is The Establishment manipulating technology that makes people greedy and unwise. When this concept is related to agriculture we enter yet another dimension of the anti-technologist's unreality. It would be fair to say that the anti-technologist also is a person who believes the world to be overpopulated and underfed. Some 90% of the world's exportable food comes from North America, the most technicized agriculture in the world. In their logic, however, we should discard North American agricultural technology and move to a more "natural" and "harmonious" production system — and a result certainly reduce our exportable surpluses to zero or even face shortages ourselves. Or do the anti-technologists envision a mass movement back to the land, to a new America full of small highly labor-intensive farms, farms whose yields could continue to be high? That this appeals to many does not make it any more realistic.

Anti-technologists claim, further, that basic and critical decisions affecting the lives of millions are being made by a small number of individuals, who control technology. These controllers, it is avowed, are beholden to no one, only to their self-interest. These controllers, it is also claimed, are a new elite, with a new technological language understood only by the elite, operating independently from the mass of people, uncaring as long as people dutifully consume those products the elite press upon them.

In relating this to technological agriculture, the thesis again becomes farcical. In 1977 there were about 3,200,000 persons permanently employed in agriculture, compared with 87,300,000 persons employed in other occupations. Slightly less than half the farm workers, some 1,600,000, were self-employed, persons one might assert to be agricultural decision-makers. Here indeed was a small percentage of people making decisions about American agriculture. Have we found, then, our technological elite for agriculture? It seems almost fatuous to ask the question, but there it is — the charges have been made repeatedly.

First, to have several million persons organized as a conspiracy, let alone several million American farmers, stirs one's sense of incredulity. To go on to imply that the conspiracy is indeed in existence and operates in concert and with enormous subtlety in its manipulations, as it must, and in secret, is plainly ridiculous.

Basically, farmers react not as a group but as individuals. As such they react usually to market prices. In turn, market prices in agriculture reflect a number of interwoven and often hostile factors. Weather at home and abroad, international tensions, trade barriers or restrictions, the economy at large, interest rates, long-range forecasts, availability and price of alternative commodities, and consumer demand are a few of the more obvious characteristics. Other non-agricultural elements of a modern economy presumably are equally complex, if not more so.

Therefore, it is suggested that the conspiracy theory is yet another piece of nonsence by anti-technologists. Clarity and light are never part of an atavistic or millenarian movement. Rather, the features are mystery, darkness and a brooding sense of oppression arising from forces hard to define and seemingly omnipotent in their power. A conspiracy theory therefore fits this mental set quite well.

In the main, our environmental atavists have idealized Nature. It is pure idealization, for they know nothing of Nature. How could they? They talk of woods, lakes, mountains, rushing rivers and the gentleness of fauna in the wild state. All this exists in nature. But so do other things — nature's cruelty, unresponsiveness, unforgivingness and indifference to man. This idealization as a further symptom of developed atavism.

One story might illustrate the point. As a young man I remember driving some cattle with a Danish cowboy. The day was bitterly cold, with sleet borne on strong winds cutting into our very marrow. The animals were miserable, breaking to seek shelter in the adjoining forest, which became not a thing of beauty (as it sometimes was) but an enormous obstacle. Horses and dogs were tired — looked homeward — and it grew colder yet. There was no harmony here with nature. Disharmony, in flooding spasms, was the theme. Oh, to escape from it! My older Danish friend summed it up as with numbed limbs we tried to take off our sodden, frozen garments when finally we reached shelter that night: "Dis nature, she is one son-of-bitch!"

Yet another characteristic of an atavistic movement displays itself among contemporary anti-technologists. They are certain they know what is right for all of us. They know that we are all unhappy and dissatisfied with our lot in a technological society. They know that even if we think we are happy we are merely deluding ouselves. We delude ourselves rather stupidly at that, for we are under the subtle manipulation of the mysterious "they" who feed us baubles and trinkets to take our minds off more important things.

The irascible ones, in particular, who are also our atavists and millenarians, feel new angers in that we are so perverse as to think life is better than it used to be, that technology may be part of the pursuit of happiness. Like all atavists, the anti-technologists have a poor opinion of the common people, those of us who can be so easily seduced. This leads them again to one of their more sinister characteristics. We who are so easily persuaded need to be made over into correct thinking persons. We must be made to see our folly, the cheapness of our minds and imperfections of our doings. Who should make us over? Why, this new elite should make us over. It is a duty. They are the new god-seekers.

Possibly what we are witnessing with extremes in the environmental movement is birth of a new religion — or should one say a very old religion, that of animism. Animism is here used in its anthropological sense, rather than that of the related doctrine emulated by Stahl in 1720. What is meant here is the attribution of a soul or spirit to inanimate natural objects and phenomena. Thus nature is accorded a god-like quality of omnipotence born of its assumed perfection. Nature becomes an absolute to be reverenced but never interfered or tampered with, let alone changed in substantial degree.

This leads to several interesting mind sets. The order of the day is stasis, a stoppage of process. Nature abhors such thinking, for it is constantly changing. Species change as they evolve, species die, new species come to life, great mountains erode, rivers build new plains and tear down old ones, climate alters over short time spans and more dramatically over long ones. Presumably even man is yet evolving. This means nothing to the new animists. Like the garden of Eden, Arcady is to be changeless and timeless.

The new animism begets dogmatic behavior, and even more, dogmatic design, as is common of any absolutist view. In the U.S. we have a very primitive animism arising among us. Not all animistic religions are this way. I have lived among Maiou, Lawa, Mussor, and Po Karen. All are Southeast Asian hill tribes and animistic to a man. But theirs was an old animism. It had a wide body of knowledge and law, even a liturgy. There was room for rationalization and thereby accommodation and compromise. The new animists, unlike this, have a primitively simple view about nature: it is absolute and must not be touched. Those who do tamper with nature must be prevented from doing so, even exorcised. If the analysis is right, we shall have many difficult times with these folk.

It is also interesting that the environmental movement in the U.S. spends less than 5% of its funds on scientific research.[14] This presupposes they have the answers a priori or they do not care about answers. The dogmatism this gives rise to is not uncommon to some religious movements.

Men do tend to be a pretty bad lot. But this has little, if anything to do with technology. Greed, hatred, despair, violence, cruelty, all existed before contemporary technology existed. Moses sought his Ten Commandments, dealing as they do with human frailty and barbarism, not because he wished to indulge himself in an academic exercise, but because his people were in sore need of admonishments and rules. Man is born but to die; death is the price of life. Along the way he suffers, in one degree or another, sickness or old age. Technology has not relieved us of all the burdens of being human and is unlikely to — ever. Technology need not be worshipped, for it is nothing but a tool. In this sense we must be careful not to maneuver ourselves into exactly the same psychic state as the anti-technologists, but on the other side.

If used intelligently, technology can help alleviate some human difficulties. More importantly, if we seriously contemplated a future where bodily succor needs to be provided for vast numbers of people, where intellectual stimulation ought to be widespread, and where we must as a species reach out for our survival to other planets, technology becomes a necessity. Man, too, is an evolving creature, as yet unfinished in his bodily, mental and spiritual dimensions. It must now be apparent, not as an article of faith, but as a simple, prudent reality, that technology is a particular dimension of that evolution. We are in an arrested evolution if a female must spend her time grubbing for roots to cook and eat, and man all his time on the vagaries of the chase. This is the true state of man in nature without technology.

By and large, anti-technologists want to retreat from the human dynamic. This rather chaotic experience, the experience of being human, upsets them. The anti-technologists should be induced to beat a retreat. Retreat, like confession, is often good for the soul. But our irascible ones want everyone to retreat. This is not going to happen. Thus their irascibility becomes greater. It still does not happen. Then formulae are devised to make us retreat. For example, political processes are invoked, where if we do not see the light voluntarily, we will be forced to be good. They wish no less than a change in human nature.

Now, changes in human nature do occur. Usually they come about very slowly. Edmund Burke noted that the only change that had any permanence is that "which occurs by insensible degree." In modern times some have tried to bring new men into being quickly. Their methods have been brutal and seemingly futile. But futility does not moderate a zealot's efforts; on the contrary, he redoubles effort. After all, the zealot is saving man from disaster. What could be more noble? For such an end any mean is justified.

Basically anti-technologists cannot stand human freedom. Human freedom is indeed a messy business and is extremely annoying to those who have visions of a new order. The anti-technologists undoubtedly have such visions. They see establishment of the new order as an essential to human survival. They are moved by powerful forces and are but a step away from becoming totalitarian in their mind-set. Will they take that step? History indicates that they will. Perhaps the current crop of anti-technologists will shrink from such a positive step, but history tells us that totalitarian dictatorships always have fore-runners. These are usually intellectuals seeking to formulate a philosophy for their own guidance while, at the same time, they preach counsels of impending doom or a doom already arrived. It is the second generation of leaders, those who decide to become saviors, who mobilize and organize the masses. By organization, action towards a particular end can take place over time and over space.

Will anti-technological fantasy reach this point? Probably! But it does not have to develop that dramatically to produce serious effects. Our democratic system is saddled with politicians whose chief motivation is political survival. These men will embrace any fad that promises them survival. This is not cynicism, but a simple political fact which we ought to recognize. Politicians seeking survival make promises which sometimes they are forced to keep. Politicians too, it must be observed, often do not understand reasonably complex technical matters very well. We have noted the punitive Federal pollution-control legislation and the dangerous dynamic introduced into the production of pesticides by stupid, short-sighted, and arbitrary regulations. Under the environmentalists' bewitching pressures of incantation and prose, we may have more. Plenty of politicians will pick up the fad if they think it will give them a platform. Others will see their chance to play the role of savior. Politicians love this role. At this stage of our political development, we should expect more anti-pollution and related legislation no better conceived or written than its predecessors. It will curtail growth in agricultural production. Thus we may reach a stage where the worst prophecies of environmentalists will then be fulfilled. Food shortages will increase as the anti-technologists said they would, and not diminish as I have asserted they can and should.

If ever the full credo of the environmental movement becomes enforced law, one can confidently predict that American agriculture as we know it will collapse. Indeed this is what environmentalists seem to want. They want to replace present methods with a kind of arcadian peasant style of production. Peasants and peasantry are important elements in this planet as we shall see in the next chapter. If properly handled, peasants can be the most productive agriculturalists of all. But the future of the U.S. lies in our technological capacity. We cannot let this future capacity be destroyed in the interests of a romantic notion.

We must be realistic. The probability of the U.S. lapsing into a peasant society is not high. A lapse into anarchy would be much more likely if attempts were ever made to curtail technology. By and large, Americans are not overawed by anti-technological arguments. They prefer to muddle along, to be as Chesterton once said, "their potty little selves." We in the U.S. are in fact interrelated with technology and particularly dependent upon it. Technology is not always good. Mistakes are made and some of these could be serious. As the anti-technologists have noted, these mistakes make us even more dependent upon technology. They deplore this and they should not. Technology augurs to remedy its own errors, quickly.

We should never be careless about atavistic movements. History shows us that they can, at the right time, attract mass followings. As of now the atavists have virtually no following among the ordinary folk who do the work and pay the bills. Their following is primarily among their own kind, among folk with their own narrow experience of life and among the quite young. A violent or dramatic change in the welfare of people, such as the terrible material stringencies that might be the aftermath of a nuclear war, or a massive economic collapse from other causes, could motivate people to look for saviors that offer them new hope — any hope. In these times people are less interested in doctrine than they are in the promise to alleviate their travail. The atavists in a primitive, involuted way might offer a hope in a distressed situation. They would offer a new idyllic existence where we would not have to worry about starving millions abroad or face practical challenges posed by burgeoning technology at home. That the terrible price of this would be a total loss of personal freedom, a stifling of all creativity and a loss of the sense of adventure of life — all this would not be noticed until it was too late. It is always this way.

The worst characteristic of the irascible ones is their lack of faith in people. They seem not to want to recognize human creativity, for this poses a problem for their beliefs. Man has powers to meet challenge, overcome and even be the arbiter of his own individual destiny. As an ultimate environmental heresy man, through his own work, may often improve mightily on Nature.

The dire warnings in this chapter are just that, warnings. Today the irascible crew is not especially important, either as prophets or shapers of destiny. But we should always cast upon them a most penetrating glance. In the meantime let us turn to folk who are important, the peasants.

FOOTNOTES

1 Chiaki Nishiyama, *Pessimism in Spite of Omnipotence: Optimism in Spite of Limitations; Comment on the Report of the Club of Rome.* Paper presented to the Fellows, Woodrow Wilson International Center for Scholars, November, 1976.

2 James A. Weber, *Grow or Die!* (New York, Arlington House, 1977), p. 170.

3 *Ibid.*, p. 172.

4 The seriousness of the situation has been recognized. Legislation is pending which could relieve the family farm from this destructive burden.

5 Norman Macrae, "United States can keep growing — and lead — if it wishes," *Smithsonian,* vol. 7, no. 4 (July 1976), p. 34.

6 Harold J. Barnett, "The Myth of Our Vanishing Resources," *Readings in Human Population Ecology,* ed. Wayne H. Davis (New Jersey, Prentice Hall Inc., 1971).

7 Arthur H. Clarke, *Report on Planet Three* (New York, Signet Books, 1972), p. 130.

8 In this study the only yields quoted are those obtained "on the farm." A normal yield of corn for example in the mid-west USA, is approximately three and a half tons per acre. In research institutes however, yields of sixteen tons of corn per acre are being reported. Similar increases in yields are reported for most other common crop plants. Under specially controlled conditions such as John R. Meyer's experiments at the University of Arizona yields of corn have been advanced to 15,400 lbs. *per day.* These figures do represent potentials, but the theme of this book is essentially reality at the farm rather than speculation upon the extraordinary possibilities that could be at hand.

9 R. Keith Walden, President, Farmers Investment Company, Arizona.

10 Most of the larger chemical companies have gone out of the pesticide trade. The profit margin is simply too low to pay regulatory costs. Monsanto now produces only one pesticide, Exxon none.

11 Lewis J. Perl, "Ecology's Missing Price Tag," *The Wall Street Journal* (Aug. 10,1976).

12 Jacque Ellul, *The Technological Society* (New York, Vintage Book, 1967).

13 For the best analysis, indeed a classic analysis, of the dynamics of atavism and its sister, millenarianism as well as a comprehensive bibliography see Norman Cohn, *Pursuit of the Millenium* 2nd Edition (New York, Harper Torch Books, 1969).

14 The remaining ninety-five percent is spent on influencing legislation and propaganda.

PEASANTS AND OTHER PEOPLE 4

The issue of planetary agriculture cannot be pursued without relating it to peasants. They are much more important than our previous actors, the irascible ones. There are more of them. Of approximately four billion people on the planet, at least two-thirds, probably more, are peasants.

A peasant lives in the country and works on the land, either as a small farmer or a laborer on a small farm. The operative word is small. A peasant is always a country man, what poets like Gray, Wordsworth or Hardy would call a rustic. A peasant has normally been thought of as a person of the lowest rank, the opposite of a person of noble birth or of the intelligentsia. This is unfair and unnecessarily degrading. Peasants are usually poor and represent, as truly as does anyone, common people. But in ancient times peasants were distinguished from serfs or other slaves for, as Seeley said in 1878, "a peasant properly so called must have a personal interest in the land."

The peasant was also often considered to be rather a boorish fellow and something of a rascal. In Europe, in particular, he was not considered as a person who had rights that deserved honor or respect. Shakespeare put it well in the "Merry Wives of Windsor,"

I will predominate over the peasant
and thou shalt lye with his wife.

But peasants have also been accorded useful accolades, if not roles, by those who considered themselves the peasant's betters. Princess Eugenie, spouse of Napoleon II, selected a handsome and strong peasant to nurse her child. In more recent times Mao Tse-tung made peasants the focal point of Chinese life, and we shall make some observations on this later.

In a seemingly short-lived neo-romantic movement in the 1960's, affluent American youths mythologized the peasant as the true son of nature. Some attempted to live like them, albeit for a short time. Tolstoy even saw God in a peasant's simplicity of existence, yet at other times he railed against peasant ignorance, sloth and superstition.

It is said that peasants have created little in an aesthetic sense, and it could generally be said that they have no literature. Some Irishmen might object to this when they think of the rich, gripping, sad and beautiful words of indigenous Irish poetry and prose. All one can do is to beg forgiveness and point out that not all peasants are Irish.

It is also said that peasants do not create history, that they are not prime movers. True as this may be, we must also note that peasants have been a great force in history. We have briefly noted millenarian and atavistic movements in Europe between the 11th and 16th centuries.

Peasants formed the mass, the inner dynamic of these movements, and we can still ask ourselves whether the deep psychic drive that pushed these people to such excesses, and ofttimes so near power, still lies hidden and can again be aroused.

It was basically peasant forces that won two victories in Vietnam against modern, technological armies: first against the French and then the Americans. When Mao Tse-tung made peasants the focal point of a Marxist revolution, he turned Marx on his head. Marx said revolution must be carried by the proletariat, itself a creation of technology. Not only did the existence of a proletariat signify a particular stage of historical development, said Marx, but its existence was vital in terms of moving to the next stage of historical development. A revolution, therefore, that adhered to Marxian dialectic — a revolution to usher in socialism — was impossible in a peasant society. It would happen only in a technological or industrial society.

In the 1920's Mao Tse-tung endeavored to follow Marx. He organized China's limited proletariat in towns and cities, but when these adherents moved into action they were crushed. Mao then turned to China's peasants and negating this important aspect of Marx, mobilized peasants and eventually won his war. Mao did not then drop his peasants. He could not; they were China. He killed millions of what to him were the wrong kind of peasants, those who had "a personal interest in land," and he endeavored to politicize the rest. Peasants were the focal point. To Mao peasant existence became so healthy, so vigorous and instructive that he sent his bureaucrats, intelligentsia and young people to work with peasants in the fields: to learn from them.

In the 1870s in Russia the same phenomenon, that of the young going to the peasant for spiritual succor, as a "pilgrimage of grace," also occurred. There was another element. As Edward Crankshaw put it,

> **Poor students and rich students, girls and boys, undergraduates and postgraduates — all turned their backs on their familiar lives and went out (to the peasants) to preach the gospel of self-help through revolution.**

The movement to revolutionize the Russian peasants failed in typical fashion for such experiments. Speaking of the student experience Crankshaw goes on to say,

> **Their own awakening was bitter. Most of them had never talked to a peasant in their lives. They were appalled and shattered by their reception at the hands of the noble savages they had come out to liberate. They were jeered at, abused, sometimes beaten and stoned, reported to the police. They discovered that the Russian peasant, to generalize, was sly, suspicious, envious, venal and drunken. They were shocked to find that the most able and articulate lived for the day when they would be rich enough to employ and exploit their weaker neighbors.[1]**

The idea persisted, this notion of peasant being a root of society...its hope of rejuvenation. In Russia shortly thereafter Dostoevsky romanced upon the "holiness" of the Russian peasant, personification indeed of the Russian soul. Tolstoy personally and physically did the same thing, thus turning to the peasant nearly three-quarters of a century earlier than did Mao. But Tolstoy also railed at peasant unresponsiveness. It may be an idea whose time is yet to come — this learning from peasants. Who knows? It might be salutary for non-Chinese and non-Russian intelligentsia and young folk, to say nothing of politicians and bureaucrats, to learn of symbiosis of plants, animals, climate and men. A peasant experience could teach them something of this.

In America's history also, peasants loomed large. It was the peasant millions who settled America, who provided most of the muscle and some of the brains. They came from all over Europe, eight million alone from tiny Ireland, eight million or more from what is now East and West Germany, millions from Russia, Poland, Romania, Hungary, Italy, the Balkans. Incredibly they, or certainly their children, all became Americans, which tells us something else about peasantry. They are not only tough and resilient; when given the right environment, they are adaptive to change.

Whatever one may think of peasants, they exist. Indeed they exist today in greater gross numbers than ever before. No view of contemporary agriculture and its future has any meaning unless peasants and peasantry arc taken into account.

There is enormous homogeneity in a village. People may be related but there is, in most cases, not too much inbreeding. Probably through empiricism these folk have learned dangers of inbreeding. Some cultural groups have elaborate systems to insure that wives are found in other villages so dispersed as to make inbreeding unlikely. The homogeneity of the village comes, rather, from living together, in relative isolation and in such small numbers that everyone sees everyone else at least once a day.

This has mixed attributes. Gossip is a normal part of village life, and one's business is everyone's business. People think nothing of asking intimate questions which in urbanized societies, even in the style of the 60's and 70's, would never be asked. Strangers are not spared, either. There is often a sense of stultifying antagonism in the village toward anyone who is different. A man who does better than his fellows for one reason or another meets strong disapproval. In some societies the fruits of his advantage will be stolen from him so that the norm is maintained and the even tenor of village life is not disrupted. To attain "upward mobility," one has to leave. There is also much backbiting and petty bickering in village life.

On the other side, the homogeneity of the village does bind it together in terms of important events. There is a strong sense of cohesion, of working together on critical things. Normally villages cooperate in planting and harvesting or such matters as digging a well. They will consult with one another on political matters, on how to handle economic hardship, on how to handle officials and their visits. They celebrate birth and marriage together. They mourn together. They will punish crime and often temper punishment with mercy if the person is young. But there is little crime.

They perform rituals together and sometimes get drunk together. A village is a tight little island set in its fields, quite self-contained and self-sufficient. But despite the romanticism of most affluent Westerners, a village is not a particularly pleasant place in which to live.

In material terms life is squalid and uncomfortable. Within weeks the visitor from outside begins to long for a comfortable chair or a reasonably modern bathroom. One begins to dread long night. Bugs falling on one's sleeping place from the thatched roof soon cease to be a novelty. In rainy seasons, one becomes sick of mud and there is no escape from it. In dry seasons, there is no escape from heat and dust. Perhaps the feature that builds the most tension is lack of privacy. Ceaseless talk and chatter, continuous prying, is something that presumably one has to be born to. But villagers complain of it also.

Nevertheless, it is from these multifarious villages, perhaps 20 or 30,000,000 of them that come the people who produce the bulk of the planet's food. Any change in the way in which food is produced in future will involve these units and their people. We often forget the scale of their production. American farmers produce more food surpluses than any other farmers in the world and more food per person, but peasants produce more food in gross quantities than anyone else. Such peasant countries as Indonesia, Thailand, South Korea, Sri Lanka, and Bangladesh all produce more rice than the U.S. Even in terms of wheat, India and China each produce about four-fifths of the U.S. crop. Only in corn, primarily for feed for animals, does the U.S. far outdistance any peasant society.

Yields in peasant societies vary: Japan, insofar as Japan falls into the category of a peasant society, produces the highest rice yields in the world, about twice that of the U.S. Taiwan and Korea also produce more rice per unit area than does the U.S. Laos and Cambodia, however, produce less than half of the U.S. yield. This pattern more or less holds good for all other major and minor crops.

In some peasant societies we find the world's highest yields; in others, the lowest. It is to the higher yields that we must turn our gaze. How does it happen that basically small farms, with minimal energy inputs beyond human energy and with limited machinery, can out-produce the nearly miraculous agriculture of the U.S.? Surely this is significant for other peasant societies. With so many peasants about, surely this is significant for the entire planet. Peasants thus must claim our attention. We must realize that peasant societies will be with us for a long time. Thus their success or failure is a critical part of world agriculture.

Peasants live in small communities; many millions of these are scattered over the planet. In their physical construct, villages vary from culture to culture. They vary in the way they are laid out and the way houses are built. They vary in terms of interiors and furnishings. They vary in the way food is cooked and eaten, and in terms of combustibles used. The people's dress is different, culture to culture. As is obvious, what might be called peasant mores, the habits, conditions and attitudes rooted in each person, also differ widely from culture to culture. Each is distinctive. But there are similarities in the way peasants live, which for our purposes require some examination.

All peasant communities are small. One can often walk from boundary to boundary in a village in a matter of minutes. The Athens of Socrates, Aristotle, Plato and Pericles was also small. It was one of Athens' assets. Athens at its widest part was little more than a mile across and about 55,000 people lived in this confine. They all knew each other more or less and there is little doubt that there was a sense of belonging to the Athenian community. A peasant village is not small in the Athenian sense of smallness. It is so small that it has no stadium, certainly not more than one place of worship, possibly not even one school of even a minimal standard, given over to elementary figuring, reading and writing, and these attributes increasingly are becoming tools whereby peasants are subject to propaganda from a central government. The typical village is so small that it cannot support a doctor or even some minimal medical service. A peasant village is lucky to have a hospital within reasonable travelling distance or a clinic which visits periodically.

Electricity is still fairly rare in villages. Indeed one feature of village life in remote areas is long nights, nights when one reads by lamp if there is one, or by flashlight when a lamp is lacking. There seems little doubt that those long nights do contribute in some measure to higher birth rates. (It should not be assumed, however, that a universal advent of electricity would significantly cut birth rates.)

The streets are rarely paved and offer dust in hot, dry periods and mud in rainy season. There are no public toilets or other forms of drainage. These facilities may not even exist in village homes. Roads from village to village are usually bad, often impassable in rainy seasons. Increasingly, though, villages are being linked one to another and to larger centers by truck and bus. These trucks and buses perform near-miraculous feats. They may carry what by normal standards is a double load of freight and passengers, while every spare space and especially the roof of the bus is laden with various peasant articles, produce and animals. As a tribute to the manufacturer, they break down but rarely. Sometimes, in some regions, villages rely on canals or rivers for communication. These provide an excellent avenue for moving heavy loads, seeds or produce.

Electronics are increasingly affecting village life. The transistor radio with its portability and long-lived battery has performed a communication revolution — and peasants took to it as ducks to water. Television fascinates them. It is becoming increasingly the vogue, even if there may be only one set per village. Nor have governments and political ideologues been slow to see the control and propaganda capabilities of electronics. In some countries the spread of electronics to villages is governmental policy. What is arising is a new village awareness and a new language, a new means of control. Illiteracy in its formal sense has been increasing, not decreasing, as population growth outstrips the often lackadaisical attempts to teach reading and writing. Peasants are learning a new electronic language of sounds and visions, the portent of which nobody really knows.

In most peasant societies there are no banks — a serious, if not crippling, situation in terms of rural development, especially agricultural development. Many peasant societies raise only one crop per year, which means one payment per year. For any kind of serious development to occur, rural credit is a necessity. This is often provided by money lenders, usually an enervating factor. Interest rates are normally very high, 30% being the average rate in peasant societies where money lenders reign. It is often much higher. Interest rates are never so high as to kill the goose laying the golden egg. Foreclosures are rare, again so as not to kill the goose. Indebtedness obviously escalates when a peasant finds himself under threat of foreclosure. It is also common to find that money lenders are aliens, often minorities. This is a security arrangement for them. They have security arising from the unity of an undefinable group often forced to be totally homogeneous because its members are alien. They are usually vigorous, and can be seen traversing the countryside tirelessly.

It might be said that in straight farming terms the high productivity of such farms as in Japan and Taiwan, comes about through four interrelating factors, all part of that skillful husbandry we noted earlier as being so essential to high-yield farming. First is a perceptive and intelligent mix of the best in modern techniques adapted to local conditions, a mix often between sophisticated and seemingly primitive. The second factor is that of taking advantage of every square inch of cultivatable soil. It is not uncommon for a Japanese farmer to grow grain, including rice, in rows. Between the rows he will plant other crops, usually vegetables. Third is an intelligent and energetic use of time. Crops are planted when they should be planted, harvested with they should be harvested, and new succeeding crops planted immediately. This seems a simple matter, but it is not. If one does not plant in time, one often does not plant at all. A cycle is lost, never to be recovered. Fourth is simply attentive and skillful husbandry. One does not find weeds in a Japanese field. Insects and rodents are controlled, and harvested foodstuffs are carefully guarded from their depredations. In India, on the other hand, it is estimated that weeds, predatory insects and rodents consume or reduce yields by fifty percent. Husbandry is a state of mind.

We are moving out of the purely productive arena into the psychic dynamics of a culture. We are now talking of results and results speak for themselves. Once I was sitting on a peak overlooking a small island in the Inland Sea of Japan. The scene before me was unbelievably beautiful: a hundred islands in a calm blue sea, pine covered. Small farms here and there, tiny villages, fishing boats and, on the horizon, ships of commerce. As I had climbed up to the peak I had noticed four or five disconnected fields terraced from the hillside. The hillside itself was decomposed granite, a harsh and forbidding soil on which Japanese grow wonderful stands of pine. The cultivated fields were different. Soil here was splendidly textured and capable of holding its percentage of water and nutrients. It had, I reflected, taken several centuries or longer to build these fields to this lush state through patient, wise and attentive husbandry. The rice and vegetables supported the idea. They were healthy, strong and handsome. Everything bespoke of good farming. Alone on the peak one could reflect on it.

A little while later a man toiled up the hill towards me. He was the farmer. He was a wiry, leathery person of about sixty years, with bright intelligent eyes. He talked of his farm. He was the kind of man who, despite his tiny holding on an isolated farm in the Inland Sea, could have talked farming with anyone as long as they, too, were farmers. The interesting thing was what he did with his hands while he talked.

There was a tiny pine tree in front of him, not more than two feet high. It was already gnarled and taking on those intricate branch shapes that only Japanese pines seem to acquire. The farmer wielded a pair of well worn pruning shears. As he talked, he clipped. He did not seem to look at or even study the little pine. A snip here, a more stubborn twig to be cut there. In a few minutes a bonsai, a natural state bonsai, was revealed. He asked me to look at the ocean through its framework. Sitting there, reflecting on the ocean through its new shape, I reflected again that a Zen master would have been pleased. The old gentleman soon wound his way back to his fields. Was this the meaning of it all, this sense of natural aesthetics? Did that too bring rich fields bearing bountiful crops?

This sense of a natural dynamic in terms of plants and soil does not come because a person is a peasant. Indeed few peasants have it. It arises from a particular tradition inherent in a culture and passed forward. It is a rare gift that we must cultivate and guard if we are to reap nature's bounty. To return to peasants and their communities, we must look at another aspect of peasant existence, this time as other people tend to see it.

There are large peasant masses in what has come to be called the "developing world" or in "Less Developed Countries," LDCs. These countries do not have a developed industrial or technological base and generally have low standards of living measured in economic terms. It is alleged that they tend to be overpopulated. But population pressures are more because of low economic levels than from too many people *per se*. People are visibly impoverished not so much because there are too many pressing against too few resources, but because productivity is pitifully low. Generally these countries lie outside of technological regions such as the Soviet Union, Western Europe, the U.S. and Canada. Synoptically they are now referred to as the Third World.

The Third World, as the name would imply, suggests a unified body or force. Indeed, it is commonly asserted that this largely peasant world is a bloc, moving forward, acting, and thinking as a bloc. The Third World has also been depicted as a force in world politics, certainly as a potential force. In the Third World, it is asserted, there are numbers, unity and a sense of purpose. The Third World is also described as being angry. Angry in its deprivation, which allegedly arises as much from the West's greedy overconsumption as it does from the Western exploitation of Third World resources. The West is warned again and again that this new, imagined power bloc is marching. The West is warned to watch out. The entire conception is a myth.

Little is unified or even common among the disparate units of the Third World. There is their peasantry, but the commonality of some aspects of peasantry must not be taken in any way to be a factor which unifies all peasants across international and cultural boundary lines. The Third World, or its peasants, have characteristics beyond peasantry just as do all humans beyond their occupational modes. In non-material sense one can ask, what of development in religion, ethics, morality, laughing and crying, love between man and woman, seeing sordid and beautiful, thinking about things, and meeting daily family needs? These things are important, no less so in peasant societies than in our own. Most humans indulge in these pursuits but they do so very differently, culture to culture, even individual to individual. These commonalities do not unite even Americans and Russians. In fact, differences are greater within the The Third World than anywhere else. Let us take but one region of the Third World, Southeast Asia, to illustrate the point.

On a human or cultural plane, Thais are as different from Indonesians as are Americans from Frenchmen, Vietnamese from Cambodians, Malays from Burmese, the enormous phalanx of hill tribes one from each other and all hill men from everybody in the lowlands. These are only some of the human strata in Southeast Asia. Neither do these differences appear only in the eye of the beholder. They are felt and cherished by the natives.

Religious differences in Southeast Asia are massive, sometimes even within the same sect. Nor does difference end there. The expression of meaning (as language says it), child-rearing, family structure, worship of objects and spirits, dress, houses, food, beverages, children's toys, education, carts, fishing patterns, festivals, entertaining, birthing, and dying; they are all refreshingly different among peoples of Southeast Asia.

These elements of difference exist in other regions of the world especially those said to be part of the Third World. One would detect massive differences throughout Africa, Latin America, India, and the Middle East. And China is different again. Throughout that vast nation there is, as there has been for three or more millennia, a semblance of a common culture. But the culture said to be Chinese is unique. And within China again there are massive differences.

Take these regions, disparate within themselves and disparate as to other regions. Do they come together? They simply do not. Why then has the notion of a Third World entity arisen? Why this conjuring up of a commonality which is said to unite men in this condition of life? Why this make-believe that a vast phalanx of the world's peoples is marching together, shoulder to shoulder? Why claim for a solidarity of political, economic and social consciousness? Surprisingly, this myth is the creation of Western intellectuals and has been taken up in varying degrees by their counterparts in the Third World. The myth is firmly believed by Western intellectuals because it fills a psychic void.

Western intellectual consciousness seems to love to have spectres haunting not only Europe, but now the West in general. Here then is another spectre. This time the spectre does not haunt Europe or even the West. It haunts the whole world. It is the spectre of a new, vital historical force, on the move to change history itself. This Third World force is to be irresistible. It represents such an enormous segment of humanity that it cannot be stopped. It is in fact unity of the greatest human mass the world has yet seen. It is a wonderful spectre.

The Western intellectual, especially the more radical intellectual, adores being frightened. Fright becomes even more titillating if the intellectual is allowed to be sympathetic with this new mighty force, even to join it. Radical thought in particular has difficulty in projecting itself over time and space, of surviving, unless it can continuously offer personal escape into other worlds. If these worlds offer to the radical a sense of being in the march of history offer the psychic therapy of escape from the miserable present to the prospect of a new and glorious future, we have ecstasy. With this adjoinment of mind to new, lofty experiences and hopes, any horror can be justified, any lie believed, any disappointment rationalized. The reasoning and rhetoric become somewhat ritualistic. Key phrases trigger common and preordained responses. A commonality of feeling and purpose arises. The world becomes cosier. It is a quite common phenomenon; it has happened before and doubtless will happen again.

In the 1930's the Western intellectual was hypnotized by the notion of the universal proletariat. He, the intellectual, might psychically (though rarely physically) even join the universal proletariat, this omnipotent historical juggernaut. In the 1930's, here again was history on the march into a world of hope and promise, a march of irresistible potency. Nearly everyone was a proletarian. Why not join, intellectually, and march out hand in hand toward the new and glorious future that was to become the result of proletarian endeavor?

It took 25 years before Lenin and Stalin's excesses and executions were accepted as "disillusionment." For some it had to be Hungary and Khruschev's denouement at the Twentieth Party Congress. All too often this shattering of the myth, this loss of historical purpose, this ripping away of unitary endeavor, left deep psychic scars on our radical intellectual. It all seemed very Homeric, except that when one is involved only intellectually, one can come to life again rather quickly and cast about for something else.

In our time the Third World is something of a substitute for that old myth of the universal proletariat. Again, one sees vast forces of history on the march. One foresees justice coming to the oppressed and feels the expiating sense of having one's own kind judged and condemned as oppressor. One feels purified by the thought that the oppressors will utimately be punished for their crimes against innocent people. The fact that most of the Third World is black, brown or yellow adds a modern dimension. Racism, as we all know, is one of the worst of the white person's assigned sins. Is there any one of us who cannot feel the relief of atonement for past sins! The Western intellectual can therefore embrace the Third World with something like the old fervor he once held for the universal proletariat.

Many will protest that some Western and Third World intellectuals are more rational, more mature than this. Indeed they are, and it is wrong to categorize so generally, but are not the rational and mature somewhat of a minority? The pattern described here is not uncommon.

As might be expected, political leadership in the Third World does not reject the Western intellectual's response. In the United Nations and in their political rhetoric generally, representatives of the Third World put on their pretense of unity. This does not make the Third World unified. Words do not paper over division and create something which is not.

The important thing is not the myth, but that this large and significant segment of humanity should be so misunderstood. If we are truly to help this great mass of people, not only in agriculture but in other areas as well, we must learn to approach them as they are. Each culture is different and sees its needs differently. As we have noted, for agriculture to be successful is not merely a matter of the practical necessities of agricultural technique. Obviously that is critical. Equally critical is the human dynamic, that element covered so far by the word "work" in our symbiosis. People do not work, and least of all do peasants, just because work is there. They work, as did our artistic Japanese tree pruner, because of an inner dynamic born of an attitude toward life. This quality is born of yet another quality... the culture to which the peasant belongs.

A Chinese child in his household learns by example and admonition that work is not only vitally necessary, but is also a duty. Perhaps this cultural attitude grew up among Chinese peasants because, as a group, they had no alternative but to work. Not to work was to die. Perhaps there are other reasons, too. The Confucian ethic which speaks of order and the duty of each stratum of society to the other, implicitly ingrained the notion of effective (working) peasantry. As the ruler had duty to people, so people had reciprocal duties to the ruler. The idea of reciprocity is key to Confucian thought and within such constraints it is inconceivable that peasants would cheat on work — at least not too much. A willingness to work is a deeply ingrained characteristic of Chinese peasants.

Not too far distant, in countries such as Laos and to a somewhat lesser degree in neighboring Cambodia, Thailand and Burma, a totally different attitude to work is found. A surplus of agricultural land and a four-month rice crop gave rise to different attitudes. It was unnecessary to work all year round to make ends meet. Enough rice could be grown in that short season to give sufficiency. Therefore why work at all — until the next rice season. Some twenty years ago I estimated that the Thai farmer averaged no more than twenty hours work per week and probably much less than that. In Laos even less work was required. Leisure indeed became a value, something indicative of success in life. Old King Sissavang Vong of Laos lamented in the 1950's, when American aid officials began pressing efficiencies on Laos,

> **Why must we do all this, why must we be pressured on all sides, when all my people want to do is eat, dance, sing, make love and be merry?**

Indeed this was what leisure meant to the Laotian peasant and to many of his neighbors in Southeast Asia.

Once again we have seen imbalances and an absence of uniform stereotypes. The differences between Lao and Chinese can be projected between other peasant groups all over the world. The image, often so prevalent in the West, that all peasants everywhere are engaged in a daily, constant backbreaking grind is simply not true. Some peasants in some regions work hard. Others, in other regions, work very little. Of the two extremes, probably underemployment is more nearly the norm than overwork.

The peasant in his work life does not necessarily have it as easy as his short working hours might indicate. His work is seasonal. He may work exhausting hours at planting and harvest times, which are crucial. If planting fails, he may not get another chance; he may face starvation, for he does not have money in the bank. Indeed, he has only a small chance of obtaining credit at high rates and he has no reserves of food. Although absence of buffers will again vary from region to region, it indicates the critical nature of peasant operations. Let us take as one example the planting of a wet rice crop in a monsoon agriculture.

The monsoon, in Southeast Asia, cannot be guaranteed, but it almost always does come. Rains normally begin in late June or early July. Prior to rains the peasant can do nothing. His fields are baked so hard that his buffalo-pulled wooden plow is powerless to break that "stubborn glebe." He must wait for nature. The rains do come. Soil in the fields begins to soften. Now the peasant must spring to action. He has a short time spectrum within which he must perform skillfully, successfully, and on time if the crop is to be planted and mature. As the ground softens he begins to plow. Also, with the now-soft soil, he mends low dykes around his field so that water may be captured. After he has plowed he scarifies the soil into muddy tilth and watches anxiously to ascertain if at the same time the field is filling with water. Ideally, young rice seedlings need about ten to twelve inches of water.[2] He gladly walks behind the buffalo, still scarifying his field, as rain pelts down and as he diverts other water channels and rivulets into the dyke-contained area.

While this operation proceeds, he must also synchronize into his agricultural plan another phenomenon. By the time the field is full of water and the soil prepared, he must also have grown out his rice seedlings so that they are the right size for planting. Previously, again selecting the time carefully, he will have prepared a seedbed and sown his rice seed in close rows. They, too, need attention, primarily watering, at the same time he is preparing the field. Now, as field preparation ends, the seedlings have grown to their right height, say eighteen inches. They are then removed from the seedbed and planted in the field by hand.

If seedlings do not grow at the right rate in the seedbed, there is potential for a tragedy. If rains come and then slacken and the fields do not fill with water, seedlings cannot be transplanted. In either case synchronization is lost. The same can occur if the peasant is slothful. Normally synchronization occurs and the crop is planted. There is then little to do until three to four months later when harvest falls due. By now the rains have gone, hopefully not too soon or too late.

Harvest again calls for an intense effort, but not under the same time constraints as planting. It is a high risk operation, where synchronization of human effort with climate variability is the key. The peasant knows the risks. If he does not pull it off, he faces destitution or even death. With his harvest in, he has security, indeed, about the same security as a technological man with a tidy bank account.

This engenders another state of mind for peasants. We are discussing rice growers, but the same attitude is common elsewhere. To the rice grower, rice in the bin is security. It is more than a mere crop that he has grown and harvested. He looks at the light brown unhusked grain in its wicker container and he sees succor for himself and family for another year. His wife, his aged parents, look at the grain and they see the same thing. This state of mind is a permanent feature of traditional agriculture. A modern technologically oriented farmer gets the same satisfaction as does his peasant cousin from fruitful harvests. But the resulting mind-set in the two persons is quite different. The technological farmer measures the harvest as a return on investment, and it is one of several alternatives in his security spectrum. He may have other assets and he certainly has access to credit. If a crop fails he has the chance to try again. He probably has other crops, if not ventures, which might balance out losses in another area. He has flexibility in terms of trying something else should he fail. In moderate degree, he has the independence not to suffer disaster from a single agricultural phenomenon. The peasant has only dependence.

As is demonstrated every year, the technological farmer will, if conditions permit, switch from crop to crop. He looks to a crop as an investment opportunity, not as his only means of survival. He will take risks if the return seems to warrant it. His mind assesses and sees opportunity about him.

The peasant looks at his agricultural situation almost in an opposite state of values. To switch from rice to something else is a major, if not critical decision. He now plays with his survival. The risk he takes is not that of losing money, and even going into debt; instead, it concerns the well-being of his family in the most basic terms. Opportunity is a distant thing, a luxury, in fact.

This fundamental difference between peasant attitudes and those of developed agriculturalists has been barely recognized by those who have tried to stimulate agricultural development among peasants. The peasant will usually listen respectfully to schemes for his betterment and be duly impressed by demonstration plots and the like. He needs more than this, however, because he sees things differently. This is not to imply that peasants will not change. They will — and do. But a different chain of evidence must be presented to them to effect change. The technological farmer sees land as an investment. The peasant sees land as his only means of survival. The peasant wants to change, if one can generalize on such a vast spectrum of humanity. But he wants to know the risks in his terms.

The questions he asks are usually quite basic and different from those of his Western counterpart. He will first ask, "Is it good to eat?" After all, he grew his crop for himself, not to be a market commodity. Any surplus that might be sold did not provide him with sufficient cash to allow for substantial changes in his diet patterns even had he wanted such. The grower, in fact, likes his particular kind of rice and not some other kinds of rice. He eats rice twice or three times a day, every day. It is his staple. He becomes accustomed to its taste. This is something else he sees in his rice bin and once satisfied about survival he wants to know, "How does it taste?"

He also asks, "If I am to grow this commodity and sell it, where do I take it, how do I get it there, and how much do I get for it?" He is newly venturing upon a market economy and he is cautious. He ought to be. He is embarking upon a fundamental change in thinking. Unless these fundamentals are answered to his satisfaction, he will stay with the status quo. Under these circumstances, who would not?

Thus our peasant world has varying, but interrelated, ramifications. Peasants are here. They exist in great numbers, forming in fact the biggest grouping of people under one occupational category on the planet. Within the category there are many variations, but all peasants are joined by the commonality of providing for themselves through agriculture, utilizing traditional or near-traditional methods. Peasants are not likely to go away. Rather, their numbers are increasing faster than those of any other group. Peasants will be with us for a long time, certainly into the next century. Our future, those of us who live in technological societies, and theirs, are intertwined. World agricultural shortfalls cannot be solved without them, and we and they must find solutions.

Interrelation of technological societies with peasants to date has been poor and sometimes disastrous. This is primarily because technological societies have never really seen peasants as a factor in the solution of the agricultural dilemma. Peasants have been seen as part of the problem. Outside of sentimental nonsense and pseudomystical hypotheses about peasant virtue and humanness as opposed to technological man's material crassness and inhumanity, peasants have not been assigned a role in the modern world by the West. On the contrary they have been looked upon either as a group that in the end will overwhelm us by their numbers and the colossal extent of their misery, or as a group that the world would be better off without. In terms of peasants and Western society, we have that chilling book by Raspail.[3] The great brown, impoverished and impatient peasant mass mobilizes in Asia, seizes some ancient ships and sets out to find succor and recompense in the tiny but affluent West, and the West capitulates.

The metaphor is plain. We are menaced by a peasant horde which is coming to overwhelm us because we neither know about them, care about them and, more importantly, are able to do anything about them. Peasants cannot be destroyed, but neither are they likely to destroy. As with the rest of us, they exist and will continue to do so to the best of their ability. But unless they change their agricultural condition, the quality of their future existence can only be exceedingly poor.

Neither can Western agriculture as a total system save them. Western agricultural methods transferred as a total system would instead destroy peasants and peasantry and create problems of lethal dimensions for peasant societies. This point, of such great importance, can be illustrated in many ways. To make the point, let us stay with our wet rice farmers in Southeast Asia.

Yields of rice per unit area in Southeast Asia are for the most part very low. While they are about one-quarter those of Japan, they are slightly better relative to American yields of rice, which are double those of Southeast Asia. The American method of large-scale rice growing could be transferred virtually overnight to the great river plains of Southeast Asia, to the Irawaddy delta in Burma, the Chao Phya plain in Thailand, the Mekong plain in eastern Cambodia, and Southern Vietnam.

The American system, as with all American agriculture, relies heavily upon machines for soil preparation, sowing and harvesting, and on aircraft for spraying herbicides and insecticides. Viewed as a system, it is really very versatile and could operate with a standard of efficiency anywhere there is a contiguous, large area of suitable rice land. With these methods, plus fertilizer, there is little doubt that Southeast Asian yields could be doubled immediately. It would be relatively simple and rapid. The reason this U.S. technique, or system, should not be applied as a system under these conditions is obvious. Yields would increase, but human cost would be unbearable.

On average, about 100 families per square mile live in the rice-growing plains. Most of these families live in small villages. Their social system is deeply rooted and strong. As a group they show cohesion and stability in facing reasonable challenges and adversities. They have developed patterns of social intercourse, celebration, cooperation, and religious observances which give some meaning and a little color to their lives. They have children, the children are inducted into the culture and there is a continuum. There are many things wrong, also. They are very poor, perhaps earning no more than $200 to $300 per year per family. Their horizons are restricted, especially as regards the outside world. Sickness and disease are serious issues. But they form viable communities and life does not present an entirely hopeless or fruitless prospect for them.

Applying the American system in their fields would double their yields. This would mean an enormous percentage boost to their existing surpluses, especially so in Thailand. The entire increase would be added to what they already sell, beyond their personal needs. If there were a market for their rice, these peasants would enjoy a great jump in their personal incomes. But they themselves, people as individuals and as communities, would not be needed.

Instead of present production modes, which utilize 100 families per square mile, we would have introduced another production mode which requires no families per square mile. No person with even a modicum of knowledge, let alone reasonably humane feelings for other persons could ever advocate such a course. Peasants in the Soviet Union were driven off the land in the 1920's and 1930's. The cost in human suffering was immense. Burmese, Thai, Cambodian and Vietnamese peasants would suffer no less. In this century peasants have also moved off the land more or less voluntarily, also in large numbers. They have sometimes been displaced by new agricultural systems not requiring people as production units.

In large degree this happened in the United States when machines were introduced, especially into Southern cotton fields which had been dependent upon hand labor. Even in the U.S., with its freedom of movement and substantial employment opportunities, the price has been heavy. It might be, as so many claim, that urban decay in some large Eastern cities is one result of this human displacement.

These large Eastern cities attracted these new migrants and the city found itself supporting them in significant numbers. It must also be admitted that cities have an attraction for country people, especially the young. One suspects that this might be the most persuasive reason for continuing migration from country to city. We are not certain. A migration which occurs where individuals voluntarily decide to move is different from one where total populations are displaced involuntarily. Voluntary movement retains some kind of human dynamic that is positive to some degree, and even creative. Forced displacement is destructive of an ongoing human dynamic and results are unpredictable, except that human trauma will be the first result.

It is estimated that there are now at least 100 million unemployed displaced peasants in the major cities of the "Third World." The real figure may be double that amount. India leads the way. We do not know in detail why these people are diplaced. Except for some countries such as Brazil, they were not displaced by modern agricultural techniques or by land exclusion policies. These masses of people probably migrated somewhat voluntarily. Population pressures on family fields, ambition, adventure, a sense of restricted opportunity in agriculture as it currently is — these might be some of the many reasons. These people have not in the main become city dwellers. Instead they have founded vast slum communities, often at a city's outskirts. As noted before, Lenin saw these people as the deracinated ones, the potential revolutionary mass. Stalin knew this also. This was one reason why Stalin maintained tight, quasi-military control of his displaced peasantry in their new encampments created to support forced industrialization.

To continue the discussion of exporting American agricultural systems, at this stage we must note not only the destructive nature of such actions in terms of people, but also the potential political impact. With one hundred million already unemployed in the cities, to add to this misery would be an irresponsible act.

We have, therefore, a dilemma — a not uncommon state in human affairs. What to do with peasants other than displacing them, and at the same time creating potential "cannon fodder" for revolutionary leaders? To date, attempts at political organizing among these displaced people have been surprisingly unproductive in political terms, but such issues are unpredictable. Probably actual insurrections must be started before large-scale mobilization of the displaced can occur. Nevertheless, the French Army during the first Vietnamese war ascertained that some 40% of the Vietminh were displaced peasants.[4] They also ascertained that there were very few true peasants — that is, working peasants — in the Vietminh ranks, other than for young boys either conscripted or seeking adventure.

At about the same time as the French were leaving Vietnam, another insurrection was coming to an end in the Philippines. The Hukbalahap movement was communist-led and substantially adhered to Marxist doctrines. After ten years of internecine struggle, the movement was suppressed by the Philippine government. Again substantial numbers of the Hukbalahap rank and file were displaced peasants, some 22% in fact.[5]

The other factor which must also be obvious is that displacement in an underdeveloped country is different from displacement in a country with a vigorous industrial base. The latter offers, in greater or lesser degree, at least a chance for a job, even if a poor job. In countries without an industrial base, displacement means unemployment under slum conditions or a scramble for such roles as shoe-shine boy, pedicab operator, or scavenger. It is better to be a peasant.

The last barrier to peasant displacement is that there is no longer an empty America. Irish peasants, eight million of them, were displaced by famine, by land exclusion practices of English colonialists, and in some degree through introduction by the latter of large-scale modern farming. The Irish fled to America and blended in. So did tens of millions of European peasants, again often displaced for similar or other unpleasant reasons. While peasants, primarily from Mexico, still enter the U.S. in large numbers, most now do so illegally. Now there is nowhere else to go: the U.S. is no longer an empty land.

Further, migration is now often prohibited at point of departure and restricted at potential entry, except for relatively trivial numbers of people. An Irish situation could no longer be relieved by the equivalent of an America. We do not know what to do with peasants. We say we have too many of them, yet peasants are farmers, and farmers produce food. We are short of food. We might pause here and reflect.

Most increases in food and fiber in the Third World (and these have been considerable) have not occurred substantially because of increased yields per unit area. They have occurred because more people are farming. Population increases in the Third World have brought about a corresponding increase in food production. Peasants are extraordinarily useful people. Relative to the planet's need for food and fiber they offer quite specific assets and even greater potentialities. The challenge is not so much of opening new lands, although this is still possible in some parts of the world. The challenge is not to export successful American production systems, because as total systems, they will create more problems than they will solve. The challenge is something new. How does the world develop a viable peasantry? Peasants know a lot about farming. Nevertheless, their methods will not do the job. They need new methodologies that increase productivity, but which also allow them to coexist with that new productivity. The challenge is to release this enormous reservoir of human potentialities. We know the potential is there because of what peasants currently achieve with so little. Where peasantry does work to its full human potential, yields are significantly greater than in the agriculturally advanced West. We have noted extraordinary yields in Japan and Taiwan. We have noted immense surpluses that could accrue if mainland China attained the same yields as Japan and Taiwan. This pattern can be multiplied many times over in many parts of the world. We can summarize that release of the human potential of the world's peasantry is the single greatest asset the world has for solving its food and fiber penury.

Another fact about peasantry: it obviously embodies the *modus operandi* of the Third World. Most of the Third World is located between the Tropics of Capricorn and Cancer. Temperatures are warm and rainfall, though normally seasonal, is abundant, except for some substantial desert areas. Sunshine, essential to the conversion of energy for plant life. is also abundant. There are difficulties in tropical agriculture. Such issues as leaching of soils and the need to learn to handle relatively poor soils are two common factors. These difficulties can be mastered by good husbandry, and tropical agriculture can be as productive as agriculture in temperate zones, if not more so.

More important, multiple cropping is feasible only in warm temperature latitudes. These agriculturally underdeveloped regions therefore have many latent assets. The natural role for these regions is agriculture and there are no unmanageable physical reasons why they should not produce agricultural abundance. Zaire, for example, located on the equator, is about as large as the U.S. east of the Mississippi. It has about 22 million people, and only 2% of Zaire is under cultivation. An adjoining country, the Sudan, is about the same size. If its water resources were harnessed and used for irrigation, Sudan could easily feed all of Africa and the Middle East. Angola has the same potential.

But yields remain low nearly everywhere in the Third World. Husbandries are retrograde. Yet, as has been demonstrated in Asia, the Asian peasant under the right circumstances can be as innovative and enterprising as the best of the world's farmers. He is not stubbornly protective of old ways if new conditions seem to give him an even chance to move forward. This author would contend that what some Asian or other peasants have demonstrated applies to all peasants everywhere. There is no racial characteristic which makes one irrevocably better than another. Here is the potential, and a truly enormous potential it is.

Now for an ultimate point about peasants. We can reiterate that few people believe in them. Significantly, leadership cadres in peasant countries often do not believe in them. If there is a typical underdeveloped country the leadership cadre is typically city-based and city-oriented. These city cadres know little of the dynamics of their own countryside. Their lives and futures are entirely city-oriented. Further, neither modernization nor the leadership group rests upon peasantry. Indeed to be a "peasant country" is to be backward. To be advanced is to have large buildings, traffic jams, international airlines carrying the country's name, factories and more factories. How then can one build one's future on peasants, let alone have faith in them as the basis and framework of a modern state?

This system perpetuates itself. The children of the city-based elites travel abroad for their education. This is a status symbol which is negated only rarely. When the young newly educated person returns, he does not see his future in the countryside, but in the city. He follows in father's footsteps as precisely as did any medieval son. If anything, thanks to his Western education, he becomes more detached, more alienated and distant from the peasant than his parents were. It becomes less a matter of not believing in the peasant and peasantry than of barely giving credit to peasantry's existence.

It is sobering to remember that in 19th century Russia, the third largest peasant society in the world, conditions were similar. Among the unsophisticated young and the unsophisticated intellectuals, there was some romanticism about peasants. This was an incidental factor. The same could be said of the new mystical qualities of godliness accorded peasantry by giants such as Dostoevsky or Tolstoy. It all did not amount to very much.

No Russian leader, neither reforming Alexander III, or the later vigorous and often gifted ministers who sought to save the autocracy, really believed in peasantry, the Russian mass. They knew it was there, omnipresent, even menacing. They took measures to alleviate the peasant condition and ultimately even allowed peasants to comment on law and procedure. But they never saw in the peasant an integral force which could have transformed Russia and perhaps saved her from subsequent miseries.

The situation is similar in the Third World today. Until the importance of the peasant is recognized in the Third World, there will be no transformation there either — unless it comes external to city elite. An external force organizing peasants could have dramatic consequences. Peasants, if led and mobilized, can move as a mighty force and when that begins no one can foretell the consquences.

To conclude this short but critical summary of peasants, it might be productive to note one situation where peasantry not only works, but offers a pattern whereby, taking account of local conditions, others can do likewise. The success of the Republic of China on Taiwan is so marked, it must be accorded a premier position regarding successful peasantry.

Taiwan is a small island, some 13,885 square miles in area, approximately 121 miles off the coast of mainland China. It is home of nearly 19 million Chinese and its population density 1,224 persons per square mile is among highest in the world. More importantly, only about 30% of the land is suitable for crops. The remainder is mountainous, although these mountains have also been put to good use. The island is subtropical in climate.

Without venturing into detail, it is difficult to give an accurate picture of modern Taiwan agriculture, except to say that in grain, hog, fruit, vegetable and fish production, their yields are either the best or among the best in the world. In fact, only Japan rivals Taiwan; it compares somewhat more favorably in fact. Yields per unit area in Taiwan in virtually every category, greatly exceed those of the U.S. More importantly, agricultural productivity in Taiwan is growing at the rate of 2.5% per year from its already high base. In addition, this small island supports an industrial base which exports far more of all products in gross value than does the entire Chinese mainland, nearly three times as much in fact. Like agriculture, the industrial base is also growing, for 1977, at about 10% per annum. It has been as high as 14% per annum. Our concern is with Chinese agriculture on Taiwan. It is in the main small scale, peasant agriculture, except for several large enclaves producing pineapples, sugar, and hogs. How did this transformation of peasant agriculture develop? How does it work?

Probably one should return to Sun Yat-sen and the revolution of 1911, when the Manchu Dynasty was overthrown. Sun Yat-sen's doctrine was simply expressed. The San Min Chu-i, or Three Principles of the People, has as its third principle the people's right to a livelihood.[6] It held that land ownership, which in the Chinese mind equated with capital, should be spread more or less equally among all peasants and land should be taxed equitably and justly. For China at the turn of the century, this was a radical concept that was never fully implemented until the 1950's in the Republic of China on Taiwan.

Meantime, on the mainland, the Kuomintang Sun Yat-sen Party split. The nation faced warlordism, Japanese invasion, and a civil war between Kuomintang and communists. But in 1948, a year before the fall of China to the communists, Kuomintang operatives in Nanking began to formulate a plan based upon the third principle which was to transform Chinese peasantry in non-communist China. The U.S., then closely working with Kuomintang, was involved in giving material and human resource assistance. The venture was launched under the Joint Commission on Rural Reconstruction, the JCRR.

Immediately after the mainland fell in 1949, the Kuomintang, now in Taiwan, reactivated the JCRR. The staff was small, a mere thirty-six members. Again preserving simplicity and directness, four operating principles were enunciated: First, the peasants must express their needs and JCRR's services must meet them. Second, there must be peasant-level agencies able and qualified to utilize effectively the assistance coming from JCRR. Third, feasibility of any project should be demonstrated before applying any of its methods. Fourth, and most important, a structure must be created which insures a fair distribution of all accrued benefits to all peasant operators involved.

In 1949 the Republic took its first critical, basic step. Land in Taiwan, now a Chinese province again, was basically landlord-owned. Rentals were exorbitant, usually one-half the crop to the landlord, with the remainder, often less interest, to the farmer. This was immediately reduced to 37.5%, adjustable in terms of productivity; this reduction, though small, gave not only relief but a sense of progress. By 1952 a land-to-the-tiller program was introduced and within a few years some 92.5% of all Taiwan farms were sold to the cultivators. This project was financed by issuance of government bonds to compensate former owners. Incidentally, this paper was in the main reinvested in the Republic's new industrial and entrepreneurial enterprises.

With private ownership of land came consolidation of scattered plots into homogeneous farms, and a long-range irrigation and waste control operation. Within less than ten years, production increased by 30%, while at the same time, costs were decreased by 20%. Waste control and irrigation operation was handled by seventeen newly established Irrigation Associations, peasant controlled with technical assistance from the JCRR. Underground water was tapped in addition to captured rain-water, and river and lake waters. Hydroelectric power production came naturally from many of the water control projects. Farming, mainly fruit and other trees, extended into the hilly region. Research began on rice strains, other plants and hogs. The Taichung variety of rice, developed in Taiwan, later became one of the strains used in producing the so-called "miracle rice." Similar production advances were made in all other crops: sweet and regular potatoes, peanuts, mushrooms, wheat, sorghum, soybean, corn, jute, flax, cotton, tea and the usual range of Chinese vegetables and fruits. Today a village market in Taiwan is a cornucopia to dazzle mind and eye. A hog industry was allied with sugar-cane cropping and pineapple plantations and began to boom. Chickens and cattle began to flourish, although yields from dairy cattle in particular are still low. Fishing and fish farming began an upward rise.

The productivity explosion, organizationally at least, rested on two key concepts: a fair distribution of return and peasant-level agencies to utilize JCRR assistance. With the advent of private ownership of farms, a basis was laid whereby fair distribution of return became possible. At the peasant level, peasant agencies able and qualified to utilize effectively assistance coming from the JCRR were created almost immediately in form of Farmers' Associations. The initiative was taken by the JCRR, but by 1953 these associations came under control of the farmers and over half the associations' income in that year came from farm sources. The associations became a farmers's vehicle, not merely a device for receiving the ideas of the JCRR. They became marketing, lending, banking and purchasing agencies. They provided storage for products and an educational service.

By 1967, one farm family of every three in Taiwan was participating in the associations' credit operations, with a 95% repayment rate. One does not have to be long in Taiwan to realize the effectiveness and strength of the associations and their fierce autonomy. That they are farmer-controlled organizations can never be doubted. Usually a manager is employed by the farmers for each association, and he is trained under the aegis of the JCRR.

At the beginning of the program, one acre of cultivated land in Taiwan supported three persons. Within ten years, one acre supported 6.1 persons. The increase in productivity continues, although obviously at a slower rate. Only in Japan is better use made of cultivated land, but though starting later, the Chinese on Taiwan will soon match — if not overtake — their Japanese neighbors.

Two further factors must be noted. These are of such significance that they point the way for others who might wish to emulate the Republic of China in their own way. The amount of money invested in the program by the state was relatively small. Over the first critical five years, the Republic's direct investment slightly exceeded one hundred million dollars. Direct United States aid was slightly more than ten million dollars. Compared with other projects in agricultural development in other parts of the world, this was a trifling amount. More importantly, investment return, (near zero for most other sponsored agricultural development programs) has usually been over 5%.

The small scale of investment and high return illustrate a critical factor basic to any agricultural development program, anywhere, at any time. The key was not the provision of money, but mobilization of human resources. Perhaps mobilization is the wrong word. Rather it was a freeing, an activation of the creative impulse of thousands of peasants. In fact, it was implementation of the ethos expounded in the first chapter of this book. It was the creation of an environment where men and women were able to pursue incentives freely. It showed the way. The Chinese have a distinct cultural asset which aided them. The inner cohesion, cooperation and loyalty of the Chinese family system transferred to the Farmers' Associations and to family work in the fields. We shall look at this later, but the key remained, that of freeing individuals to utilize their latent capabilities.

The second factor is of equal importance. The original impetus came from the state, if not from the Sun Yat-sen doctrine. The state initially controlled the planning and operations on all levels quite tightly. In such situations the natural tendency is for the state to extend its control. A bureaucracy arises and grows by leaps and bounds, regulating this and directing that. The bureaucracy becomes more centralized and more out of touch with reality in field and factory. In the extreme, the bureaucracy stifles the very energies it set out to promote. On Taiwan, however, the pattern was different. As the program moved from success to success, the state proportionately withdrew. Today the role of the JCRR as a control agency is minimal. Thus a second lesson arises: as soon as possible in any agricultural development program, the state should get out and leave farmers and fields to each other.

Statistics and like pronouncements are useful in describing a program as a program. However, it is the human dimension that counts. Agricultural development is for people, not for planners or technicians or bureaucrats. What of the people involved in this success? Let us look at a typical farm and its operation.

The area of land is small, slightly more than five acres. Two families live on this small plot, a son and his family, parents, old but vigorous, with a widowed daughter. Nine people ranging in age from six to about 65 comprise the farm populace. During the 4-month rice season, the entire acreage is given over to that grain except for small plots of vegetables for home use. Rice yield totals about ten to twelve tons. This provides sufficient rice for the family, plus a marketable surplus of approximately 8.5 to 10.5 tons. Here alone is a tidy peasant income.

Immediately after rice harvest a new operation begins. Bamboo poles are wired together to provide frames for nine or 10 small houses. These frames are then enclosed by a thatch from the harvested rice straw. Soon fields are dotted with these tidy looking little huts, each perhaps 25 yards long and 10 yards wide. Each hut's interior is lined with plastic to control humidity. Five tiers of wooden trays are then installed in the inner space, leaving just enough room for people to move about. The trays are covered with soil and mushroom spores are planted. The entire construction operation, from the end of the rice harvest until this stage, has taken less than two weeks.

Mushrooms appear within another three to four weeks. The beds are picked two and sometimes three times a day. Mushrooms grow at an astonishing rate and in great profusion in these warm, humid, dark interiors. Everyone harvests, grandmother and children included. Mushrooms are stacked in wicker baskets which, when full, are strapped to the rear of a new, shiny motorcycle and carted off to the local Farmers' Association.

This Farmers' Association, one of several in the region, has contracted on the farmers' behalf with a major United States food processing corporation, which, to utilize this production source, has built a large bottling factory in the region. Staff at the Association wash, weigh, record and then disburse the mushrooms to the bottlers.

The entire process continues day after day (there are no weekends) for about eight months, that is, until the next rice season. During this period there is other farm activity. Vegetables are grown in vacant areas between buildings and the surplus sold in a nearby city which is industrializing at a rapid pace. There are chickens, some ducks (what Chinese farm does not have ducks), and five pedigreed Landrace hogs. Supplemental feed is bought for the hogs, but with the compatibility these animals have with human eating patterns, they also consume all left-overs from each family's kitchen and garden waste. The kitchens are also worth a mention — each produces the best in Chinese home cuisine.

With onset of the new rice season, huts are taken down, thatch plowed back into the soil, and fields are prepared for new rice planting. Bamboo poles are stacked away and wiring coiled neatly for the onset of yet another mushroom season four months hence.

The peasant families involved live well with a total income close to $10,000 annually. Their diet is nourishing and in such a variety as only Chinese cuisine can provide. There are motorcycles, color televisions, washing machines, sewing machines and refrigerators, all bought at discount from the Farmers' Association. Everyone is healthy and full of that unmistakable vigor that permeates all successful enterprises. The human dimension of Taiwan's agricultural development seems sound, if not inspiring. One asks, "If here, why not there?" One asks also, where is the key that can transform peasantry from a kind of misery to be endured into a productive symbiosis that gives satisfaction to individual humans?

To meet world's needs, such a transformation must take place almost worldwide. It must involve peasants, who form the great percentage of mankind. To find the key — here is the question and that is a purpose of this book.

FOOTNOTES

1 Edward Crankshaw, *The Shadow of the Winter Palace* (New York, The Viking Press, 1976), p. 259.

2 More recent research indicates that this in fact may not be true. Appropriate irrigation using much less water might do equally well. The water covering the field does, however, keep the weeds down.

3 Jean Raspail, *The Camp of the Saints* (New York, Scribner's, 1975).

4 In conversation with French officers: Vietnam, 1954.

5 Luis Taruc, *He Who Rides the Tiger: The Story of an Asian Guerrilla Leader* (London, Chapman, 1976).

6 The other two principles were: the people's right to self-determination; and the people's right to participation in the government.

A FRAMEWORK FOR THE AGRICULTURAL SYMBIOSIS 5

We have the land. We have the people. But it doesn't work. The great, near-flat plains, grass laden or forest covered, both north and south of the Amazon, are virtually untouched even by ranching. They represent an agricultural treasure-house awaiting an agricultural dynamic. Tens and tens of millions of acres of land south of the Sahara, more or less closed to humans by the river-blindness disease, await an agricultural dynamic. Further south, such a nation as Angola has been summarily described as "twice as big as Texas and twice as rich," yet most of its people live on a subsistence economy.

The Sudan, that vast region overflowing with potential, is desperately poor. The White Nile, because of geographic factors, does not flow freely for all of its traverse. In the Sudan it spills over great areas and forms massive swamps... useless water in a region short of water. One estimate has it that the Sudan, with its combination of favorable climate, abundant but unused water, good soils, could if properly handled, feed countless millions.

There are a hundred million acres in the gigantic plains of the Brahmaputra, the Ganges, and the Indus. These great rivers are fed from enormous water sheds, the world's greatest — those of the Himalayas and the Hindu Kush. There is water to irrigate on a vast, almost unimaginable scale. In comparison, the entire irrigated land of the State of Arizona is a mere one million acres, and for the entire U.S. about 40 million acres. If this vast region on the Indian sub-continent were irrigated and farmed properly, its present production could be so increased that this region alone could yield up to three-quarters of world cereal grain needs.

The drought in arid areas of the Sahel disturbed the world in the early '70s as its human inhabitants suffered savage trauma for more than two years. However, the French Ministry of Cooperatives has concluded that underground water and rivers of the Sahel could be effectively harnessed to provide irrigation for up to six million acres of cropland. Not only that — assuming normal yields, investment of some $10,000 per acre over next 75 years could produce a 6% annual return. This is obviously a highly hypothesized estimate, but it illustrates potential.

Less than 40% of the world's arable lands are being used. Even that estimate may be too low. It is difficult to assess what is and what is not arable. An Arizona desert region which in 1950 supported one breeding cow per 150 acres, under irrigation now produces three bales of cotton and 25 bushels of wheat per acre yearly.

A New Zealand hillside, denuded and eroding, produced in the 1950's, under aerial top-dressing and seeding, wonderful stands of clover and rye grass and carried two to four sheep to the acre, often more. In Israel proper use of water (it was as simple as that) increased yields by eight times in less than 20 years. No matter how one measures it, the world has large areas of land, probably more than we farm now, awaiting the agricultural dynamic.

But new lands are less important than improvement of yields on lands already under cultivation. We have noted many examples of the truly enormous advances that may be had in yield, through better husbandry and development of improved plants and animals. Improvement of yield ought to be the primary concern. To talk of developing new lands and improving yields is really begging the questions in terms of evolving a workable pattern of agricultural development. We have explored enough to know that while the physical dimension of plants, animals, soil, and climate are basic ingredients, more is involved. Productivity is a human event. Whether it is deficient or abundant, the human factors are critical. Agriculture as it has been taught and expounded has been viewed as a technique. That is but part of the equation. The key element is human interaction. This now has to be delineated and explored. First, we must pause and see if the necessary agricultural dynamic can be conceptualized. This we must do before we can intelligently explore why the agricultural dynamic is not now being realized.

We earlier characterized agriculture as a symbiosis, noting the interrelationship between plants, animals, soil, climate and work as the basis upon which food and fiber production is founded. When we talk of transferring the symbiosis, in one form or another, to another place and to other people, we must add the factor of transference itself. Historically the process has been slow, laborious and almost fortuitous. There has been no discernible improvement these past thirty years when conscious (as opposed to incidental) efforts have been made to transfer agricultural skills.

Where conscious effort has been attempted there has been considerable risk of money and human resources, the latter primarily through frustration and the dashing of high hopes. It must be emphasized that as of now we are considering only the issues of transference itself: what can be done and how it ought to be done; establishing with its own autonomy an agricultural dynamic, the purpose of which is to produce more. As we shall see in succeeding chapters, there are five powerful external factors which make transfer difficult. These five inhibitors have little to do with agriculture *per se*. Our concern now is to look at agriculture directly, to establish a conceptual framework for moving from low to high agricultural production. After we have viewed the five inhibitors, we shall establish a *modus operandi.*

It must be assumed that what is to be transferred already embraces the best and the most appropriate in developed techniques for improving plants, animals and soil, irrigation and husbandry research. We can assume that the peasant recipients will readily understand an improved plant or animal and will grasp improved methods of husbandry. Experience has shown them willing to understand new situations if they think it will work in their situation and that they, personally, will benefit. Further, the absence of advanced agricultural systems is often as asset. Despite the West's contrary opinions, peasants are often surprisingly flexible and open about new methods and devices. But Western knowledge of the transference system is quite sparse, since such efforts have usually been made not on behalf of the peasant, but on behalf of the Westerner. Western systems are offered intact instead of being adapted to local needs. Along these lines we can repeat a point already noted. Transference of Western agricultural systems, and especially of American systems, is largely irrelevant and would probably be destructive of cultural and political patterns. Nevertheless, Western technology is vital to any improvement of peasant agriculture. Technology, however, must be introduced in a new way, and this leads to the theory of "mix."

The idea of mix is very simple. We have noted the extraordinarily high yields of Japanese rice farming. This was achieved by "mix." The Japanese, for example, mixed the finest in plant biology with such traditional methods as using human feces as fertilizer. The concept of such a mix is hard for the Western mind to adopt. The Westerner creates a plant biology laboratory with modern equipment, white-coated technicians and learned researchers, and believes modernization is on the way. To him, modernization means a new pattern of doing everything, and concomitantly, a wholesale scrapping of the old. Step by step, all is modernized. It is a comprehensive process ending in total technical transformation. Furthermore, the Westerner becomes ecstatic about the visible throwing away of the old. To him, it proves that modernization is not only under way, but is achieving valuable ends.

On the other hand, the idea of mix would use whatever is practical in a given situation regardless of whether it is traditional to that situation or is a rather exotic foreign importation. To stimulate the imagination, let us use an extreme illustration: relating cybernetics to peasant cultures. For this purpose cybernetics can be defined as automated production controlled by computers. One of the many elements in cybernetics thus defined is that very little human labor is required.

Let us assume that rice growing is the economic base of a culturally viable society, and that the effort is to improve rice yield. It is not viable, however, to uproot rice growers by introducing methods which increase productivity but reduce or eliminate rice growers. In our model our advanced rice culture will need machines and other manufactured products. These should be devised to meet the needs of the agricultural symbiosis that fits a particular cultural situation and to cater to people's comforts. Let us devise a system for Laos, which, prior to its conquest by the communists in 1975, was so unique in its abilities to work and play that it allows one's imagination to run the full gamut. It is a system which retains rice growing as a culture but which cybernates industrial and other development.

In Laos there are no Horatio Alger stories in which the young man, by diligent application, and above all, by hard work for 20 to 30 years, makes his way to the top. In Laotian folk tales, specifics for success would be quite different. Success would come through magic, the fates, manipulation or sleight of hand. The idea of obtaining an ultimate success by hard work simply does not make sense either imaginatively or as a lesson of history. It is the wrong way to go about things, this hard and unremitting grind of daily labor. But perhaps hard work might not be relevant to cybernated production. Basically, cybernetics as we have used the term fits a populace that knows how to do nothing and values nothing or, at least, very little.

Thus, imaginatively, cybernetic production could fit Laos admirably. It could produce machines and products needed to intensify rice production and at the same time make possible new products to satisfy new demands arising from improved living standards. Except for the Vietnam war, Laos has escaped all trauma of industrialization. It would indeed be traumatic for Laotians to embark upon highly disciplined and regimented work as common to early and modern industrial processes. In a cybernetic production system, almost all factors in the Western work ethic are not needed — theoretically, anyway. Imaginatively, therefore, Laos, often said to be the most "backward" country in Asia, might be able to handle cybernetics better than the industrial West.

Laotians would not be bothered by disposing of the work ethic — indeed, they do not need to dispose of it for they do not have one. Laotian peasants knew how to fill spare time between critical work periods in rice cultivation. Singing, dancing, gambling, fishing, cockfighting, whoring, a long list of semi-religious celebrations, religion *per se*, and community — these were his preoccupations.

Of course this is an exercise in imagination. Cybernated industrialization is unlikely to be introduced into Laos. New communist masters there have their own ideas on agricultural development. These are probably not too different from those advocated by Western theorists. It will be recalled that the exorcising of the traditional society and the little community is as firmly adhered to by communist theoreticians as it is by most Western exponents of economic development. To modernize, one must agonize through destruction of one's traditional society, and endure the rigors of "take-off." Then comes the disciplined drive to economic maturity and some time, presumably, the joys that are said to come from mass consumption. By that time, however, little Laos of the past would be no more. But to mix cybernation of industrial production with an advanced rice culture — why not? Especially if rice culture and all its related activity is preserved and made dominant.

This vision of a relationship between cybernetic industrial production and agriculture is unlikely to materialize. But other forms of mix, all less spectacular, are necessities in spreading modern agriculture to peasants. For example, one might despair of clearing land with machetes and turn to bulldozers and other heavy equipment. But abandoning machetes in favor of chain saws could be more economical in terms of preserving lumber and could engage the primitive forester in a new production mode where, instead of being largely supplanted by machines, he becomes effective, practical, and feels at home. There is no reason why an oxcart should not have rubber wheels or for that matter why oxcarts should go out of business. In a particular situation an oxcart might be a sensible way to transport a crop, even if that crop is the result of the most advanced industrial technology.

Great irrigation systems developed in Western U.S. fit great farming vistas, not a densely populated, village-studded peasant environment. Such irrigation is usually a regional affair and could not always respect land and village boundaries. There are systems, however, such as drip irrigation, that are intimate in their constraint and can adapt splendidly to the most chaotic patterns of peasant land tenure and boundaries. It calls for a new stretch of the imagination.

It also calls for a restructuring of the symbiosis, one which retains all the conceptual components of the agricultural symbiosis yet mixes components which fit a particular situation. Above all, we do not need to throw out the old and seemingly primitive just because it is old and seemingly primitive. New elements and relationships need to be introduced into the symbiosis.

Rural credit is vital to all agriculture, especially to peasant agriculture. The absence of efficient rural credit systems and reasonably cheap money has been one of the great failures of agriculture in the Third World. There will simply be no significant agricultural development until this is put right. A rural credit system for peasant loans, because there are often so many of them per given area, almost demands that computers be used. But the current etiology of "development" tends to reject even such a concept, let alone its initiation. Whoever heard of mixing computers with peasantry? About all Western theorists would let the underdeveloped escape (in their theories) is the age of steam. Everything else, they mostly believe, has to be a step-by-step facsimile of the Western experience. We should mix whatever fits, no matter how sophisticated, no matter how primitive.

Perhaps the most important element in the theory of mix is not technical practicality, but psychological and political practicality. We should have learned by now that destruction, especially rapid destruction of one's past, brings on great trauma. From this trauma can arise counter-destructive impulses which negate improvement. Agriculture is especially vulnerable to this kind of situation.

Burke put it well:

> We must all obey the great law of change. It is the most powerful law of

nature, and the meaning perhaps of its conservation. All we can do, and that human wisdom can do, is to provide that the change shall proceed by insensible degrees. This has all the benefits which may be in change without any of the inconvenience of mutation. This mode will, on the one hand, prevent the unfixing of old interests at once, a thing which is apt to breed a black and sullen discontent in those who are at once dispossessed of all their influence and consideration. This gradual course, on the other hand, will prevent men, long under depression, from being intoxicated with a large draught of new power, which they always abuse with a licentious insolence.

It has often been difficult for Western man in modern times to accept the notion that change of itself might be not only painful, but destructive. The cry has usually been for change now, with no counting of the price. In fact, the West seems to believe that change has no price because change is invariably good of itself. This motion is a cultural bias, especially in the U. S. Here, it is anathema to believe that change, if it is to be absorbed and thereby become effective, should be relatively insignificant in its nature, or in Burke's words, proceed "by insensible degrees." To think that human social and cultural organisms, especially long established ones, have an extremely limited capacity to change at all, let alone rapidly, is almost regarded as a vote of no confidence in mankind.

The American attitude is understandable when we remember the shortness of America's past. America in the 18th and 19th centuries was a swelling dynamic continent. A kind of rapid change born of virgin lands and massive untouched resources was the essence of things. Perhaps the real American revolution was less political than material.

Individuals or groups caught up in the unparalleled opportunity offered by a new continent only very thinly occupied by human beings, had unique experiences. These persons did step from an old world into a new one and dramatic change seemed wholly beneficial. We have been suffering ever since from this rapid but incidental phenomenon. For, in the American experience, change — swift and great — was not only exciting and exhilarating; it seemed good. The American experience was unique. It will never be repeated, for there are no more empty continents to swarm over like wandering, curious children exploring a new, well equipped playground. In peasant societies such Elysian fields hardly exist, except perhaps as atavistic memories. And such are not part of the drive to provide adequate food, clothing and shelter. That is a concern of the here and now.

Historical and anthropological evidences of the impact of change on humans is becoming more clear. Change, beyond that "proceeding by insensible degrees" can induce massive and destructive instability in individuals and cultural groups. Concurrently comes a feeling of insecurity born of a loss of control over the events shaping one's present and future. One no longer knows where one is. We should reflect on those marvelous lines in Saul Bellow's "Mr. Sammler's Planet." Mr. Sammler, old in ways of the world, is talking to his daughter of change and its destabilizing influence.

> **...it is sometimes necessary to repeat what all know. All mapmakers should place the Mississippi River in the same location, and avoid originality. It may be boring, but one has to know where he is.[1]**

Under too rapid change the map is no longer there.

Along with insecurity born of instability, born of a too-rapid rate of change, also comes very often a hunger for faith. Faith can be religious. Today, however, it is more likely to be political. Faith that a political system will put things right and heal the trauma gives man a new sense of knowing not only where he is, but where he might go and how to get there. Marxism presents a ready formula in this regard. It shows man where he stands at a particular time and that the perilous plight in which he may find himself is not his fault, but the fault of great historical forces. He comes to believe that the forces are still at work, shaping a new destiny for the individual who, for the first time since his subjection to insecurity and instability, can see his future, can even see a goal.

People in this state of mind are unlikely to tolerate what they perceive to be obstruction. Nor do they accept caution and least of all, Burke's proceeding by insensible degrees. They become zealous, fervent, excited, hot-tempered, fanatical. They form groups with others possessing similar characteristics. All dissatisfactions of the past, including personal inabilities, liabilities and failures tend to be absorbed and re-focused by what now becomes a movement. When we talk of rapid change in traditional societies, we play with fire.

Change, therefore, has its drama. It can be an unsettling, security-destroying situation. It can become a revolutionary situation in the sense of revolution as a political movement determined to bring some kind of different order. Change in an extreme form, as an inducer or a revolutionary psychosis, may be good — for those who like revolutions. This notion of change is far from the popular Western ideal of change. The popular view is to see change more as a benign experience usually making material situations better and people a little happier. Change in this sense has become a pastime of the bourgeoisie, as with changing today's automobile for a later model. Yet history, including Western history, shows nothing of the sort. Change of the right kind is necessary to human progress. It can, however, produce blood, pain and danger, with no guarantee that the human condition will be improved.

The peasantry, however, that great inarticulate mass which holds the future of the planet's food and fiber production in its hands, must change its mode of production. The very basic elements of its agricultural symbiosis must undergo dramatic change. Significant as this may be in terms of potential trauma, it also holds an ameliorating factor. The results of change — better crops, better animals, better methods —

will produce new wealth, and in turn, reward peasants personally instead of rewarding the abstract state. How change is effected becomes as important as what change produces. It can be taken as a guide that results will be produced. How sensibly any given peasant group goes about reordering its lives is another matter. Assuming that the group prefers not to create an unpredictable, revolutionary mass among the world peasantry, it must examine other, more sensible ways of effecting change.

Over all, there are four distinct, interrelated phases in promoting change without revolution. Within each phase are critical, if not vital, elements long neglected in most attempts to improve peasant livelihood and, thereby, the world's agriculture. Most such attempts have been too feeble, puerile, and unperceptive to disturb the traditional peasant equilibrium.

The West feels guilt about the colonial past, its material wealth while peasantry has so little, its over-consumption (supposedly at the expense of the Third World), and charges that its own multi-national corporations are exploiting the world. So it wants to help. That much of this guilt feeling is irrational, matters not. That it is irrational is clear from the facts.

As to the colonial past, the cases of the Dutch in Indonesia and French in Southeast Asia clearly demonstrate that despite strong colonial controls, not much basic change was produced in the life-ways or productive systems of colonial populations. If such change had occurred, there is little trace of it left today; former colonial masters cannot be blamed for the reversion to much lower levels of production today than pertained in the colonial past.

Nor can multi-national corporations be correctly characterized as exploitative in the Third World. Does a corporation's creation of an industrial labor force in Third World nations inhibit agriculture, which is still bound to its age-old forms? Could those corporations long survive if they engaged in "cheating" trade practices in nations surging toward total equality? And if they could, how would this affect agriculture bound to traditional methods? As to guilt in the West about its wealth in the face of Third World poverty, surely this is misplaced, for mere guilt will not bring about any change in that situation. Only positive approaches can do this. What positive approaches are necessary and feasible?

First, native culture must view honestly its merits and its shortcomings, and it must generate the will to change itself and develop. Secondly, and more important, in this framework of cultural honesty and realism, there must be developed a widespread growth of individual will to change. Such human motivation cannot be produced by merely advocating it, or through ideological emphasis. It must result from material changes. More land should be made available for individual ownership, but this is difficult because land is often a limited resource. More important, leadership must be developed at local levels. Here foreign leadership can play a limited part. Most foreigners are not qualified for direct local leadership. They are generally too ignorant of local culture, are not language-qualified, and generally lack the necessary cultural identification.

Some may be tempted in this respect to substitute coercion for the development of individual will to change, since such development must be slow. But coercion of the peasantry, as we can see from Soviet and Chinese communist cases, is largely self-defeating. Finally, there is money as an inducement to the individual to change. It can help, but only if the peasant is convinced that awards from increased production will be fairly shared. Unfortunately this is rarely so. Although such increased sharing was present even under colonialism, it has, for the most part, not survived. The new independent state, as landlord and tax collector, has usually deprived the peasant much more than previous colonial regimes. Indirect taxes, whereby the peasant supports the urban sector in comparative luxury, are the curse of his life. Although at the outset of some revolutionary regimes, the peasant shared more than before, this has usually been short-lived, partly because new regimes have augmented their power at the expense of developing any autonomy for the individual peasant. Fuller sharing and individual peasant autonomy must necessarily go hand in hand. Maoism was defeated on this account: without augmenting peasant autonomy there can be little increased production nor any sustaining of what minor increases are induced.

Finally, for any widespread growth of individual will to change, there must be an effective supporting agency. To start, it must have cultural identification with the peasantry. This almost automatically rules out much foreigner participation since, except for some missionaries, foreigners are generally ignorant of the culture. Some even come to foreign countries seeking psychic aid for themselves in the name of "foreign aid." At best they can work only in supporting roles; primary actors in sponsorship must be indigenous and thus culturally knowledgeable. It is absolutely necessary that peasants themselves play a part in the sponsoring agency, although the gap between them and the usual government agency bureaucrats makes this extremely difficult.

Here again we run into that wide gap between country and city, normal to almost all Third World nations. The central government is almost forced to initiate such agencies of sponsorship, no matter what their concern. In fact, it has a primary responsibility here. What has gravely inhibited its work is its role of ruler over peasantry, instead of what it should be, namely their servant.

Here two principles must eventually govern, if government is to provide leadership for the will of the masses to change their traditional cultures in the direction of agricultural growth.

First, government must accept the rule of subsidiarity, that is must serve and not rule, that its concerns must be subordinated to those of peasants. Along with this is the principle that the role of government must be progressively diminished. Only with such individual autonomy can Third World peasant agriculture escape from a marginal subsistence. The key is to release individual productivity through fewer government controls.

To all this there are formidable obstacles. These inhibitors are not primarily agricultural. Nonetheless, by their infringement upon the individual creativity and individual freedom that must accompany such creativity, they are effective obstacles to agricultural development. They operate individually and collectively. What are they, and how do they operate? Let us now proceed from the general to the specific and explore the real inhibitors in some detail.

FOOTNOTE

1 Saul Bellow, *Mr. Sammler's Planet* (New York; The Viking Press, 1969).

AN INHIBITOR: INTERNAL INSTABILITIES AND INTERNATIONAL TENSIONS 6

Over slow-moving geologic time, the geography of the planet took on its present shape of oceans, continents, and islands. Approximately 70% of the planet's surface is sea; remainder, land. That there is this large body of land juxtaposed to an even larger body of water and that both are heated by the same sun is a basis of climate. Differential rates of heating and cooling of land or water impinges upon a thin film of warm air and cool air, moist air and dry air. As a result, the air begins to move. A geostrophic wind force born of the planet's orderly rotation is also added. These combined factors form the basis of weather phenomena, which inter-relate with land forms and become a part of the agricultural symbiosis. Land form and climate exist regardless of the territorial distribution of men.

Thus agriculture, which is dependent on the sum of these factors, extends beyond man's nationalisms and quarrels. The soils on either side of the border between the two Irelands, between Israel and its neighbors, between many African and South American states, are contiguous — so is the climate; so are plants and animals. It is people and the way they behave that are different. On the other hand there are climatic and soil regions of great diversity on this planet, and therefore a diversity of animals and plants.

The boundaries of these natural diversities (natural boundaries) have little relationship to man-made boundaries. In fact, we live in what might be called a cross-boundary world. For various reasons, mainly political and national, men ignore natural boundaries...they cross them, cover them up, twist them into new shapes "with licentious insolence." In the less contemporary world the same thing was done often for greed, an honest greed shown by conquerors.

Today the process is more subtle, but the directing passions are not all that different. Man manifests himself still by arrogant disregard for what nature has wrought. These are the least of his concerns. Agriculture in its symbiosis may be politically neutral, but it is not geographically neutral. On the contrary, it offers its true bounty only when its geographical constant is taken into account.

Ideally agriculture should be a planetary system. If it cannot, it should be a regional system, with a regional boundary being determined less by national argument than by ecological accord. What can accrue when the ecological accord of a region is intelligently related to man-made political and related constraints can perhaps best be illustrated in the United States.

The continental or near regional agricultural system of the U.S. is part design and part accident. Assignment of the word accident is significant, if for no other reason than that the American colonizing experience will never happen again on earth, short of a recovery from some planetary disaster.

Neither is the U.S. system entirely regional. To the north, Canada insists on its national integrity. To the south, Mexico insists upon its integrity, and so do the Central American states. United States does have trade and farm labor relations with these countries. But productivity of North America could be strikingly improved if it were a truly continental system. Let us return to the partly continental system of the U.S. to note how such a system contributes to agricultural success.

Currently 90% of all world food exports come from the United States and Canada. In 1975, 60% of all wheat grown in the U.S., some 70 million tons, was available for foreign buyers. With rice, 60% of the world's exports, in excess of two million metric tons, was available for export. Soybeans amounted to 50% of world exports or 20 million metric tons; sorghum, 25% or 10 million metric tons; corn, 20% or 60 million metric tons.

At the same time, in this continental system, the American consumer, regardless of where he was located, also benefited. In terms of private consumption expenditures Americans on average spend some 17% on food, beverages, and tobacco. In the United Kingdom, it is 32%; in France, 24%; in West Germany, 27%, and in the U.S.S.R., some 50%. And the productivity of this American system increases. In 1945, one American farm worker fed 15 people; in 1957 this increased to 23 people; by 1976 one farm worker fed 56 people.

One reason for these extraordinary figures is that the U.S. has created an economic, political and bureaucratic environment that has allowed the agricultural symbiosis to take advantage of a continent. This productivity did not arise by accident. Many observers point out that the U.S. lies mostly in a favorable temperature zone, but so do many other territories where comparable productivity does not exist. Also, U.S. has climatic disadvantages. The mid-northern states endure long snow-clad winters. In the Southwest there are vast desert regions. Yet in Minnesota, in mid-winter with several feet of snow on the ground outside, one can eat fresh lettuce and ripe tomatoes. Nature's bounty or intransigence is not the reason for U.S. agricultural success, neither would it be reason for failure. Many other factors exert greater influences. Most relate to American development of a continental system.

First, the nation has integrated its agriculture and can take advantage of climatic and soil variations that are inherent in any great land mass. When snow is falling in New York, avocados, asparagus, tomatoes, greens and other products associated with warm climes are being harvested in southern California. In April, when Iowa is beginning to dig out from its winter snows, calves are being dropped in Florida. Linking these elements together is an efficient, speedy continental transportation system. In December a truck drops a refrigerated trailer into a Calfornia field where lettuce is being harvested. The trailer is refrigerated even in the field, and the lettuce is placed immediately in this portable cold-storage system. The truck returns, hitches to the trailer and three days later, the product reaches New York, three thousand miles away.

Arizona cuts green alfalfa in a 110°F. temperature in late summer to feed a young bovine calved in mid-winter in Florida, again transported by the ubiquitous trucker. The impact of this interlinking is more than providing Americans with summer treats in winter or feeding desert-lodged calves with green fodder. First, such a system lends itself not only to fostering political stability, but also to political sophistication. Instead of infighting over who gets what share of a shortage or a long-kept inferior product, the obvious incentive is to cooperate in production and distribution. To meet market demands of a continental system is in everyone's advantage. To erect trade barriers to protect a market or a supplier makes little sense, although such barriers are suggested from time to time.

There is really only one true continental agricultural system, that of the U.S. Western Europe, despite distinct moves of the Common Market toward a continental system, still faces barriers to free intercourse, especially in the form of national subsidies. India ought to have a continental system, but agricultural retardation, poor transportation, government political interferences and inefficient bureaucracy are hindrances. Throughout South America and in Africa there is a mixture of free trade and trade restriction, with the latter dominating. In these vast regions a continental system of agricultural distribution does not exist at all; even the agricultural base for it is nonexistent.

For example, the long and severe drought in the Sahel region of Africa in the early 1970's was recognized by the entire world. But it was not African food that was distributed to the drought-ridden region. Grains and other products came primarily from Europe and to a lesser degree from the U.S. The agricultural base of Africa was simply so low that surplus food was not available. If it had been available, the "system" was so chaotic that it would have been hard to gather it up, let alone distribute it.

There was, however, a more sinister reason why Africa could not meet what ought to have been a minor problem for a continent so richly endowed. Angola is a significant country in Africa. With 481,351 square miles, it is one of the world's larger nations. Its untapped natural riches are near legendary. But Angola as of 1974 had only 5,200 miles of paved road, compared with Texas, one-half as large, which has 67,000 miles of highway and 165,000 miles of other paved roads. In every other comparison like disparities exist.

Throughout the 1960's and culminating in the '70's, Angola was in a state of internecine warfare. First nationalists against Portuguese colonialists, and then as a culminator, a civil war between three different nationalist groups. Despite its intercontinental link, Angola took little interest in the drought-ridden Sahel. Internal tensions were dominant in Angola and the trauma of Sahel was incidental. The U.S. also has droughts and other agricultural disasters, but because of drought or flood, Wisconsin butter is not denied Minnesota, or Kansas wheat denied to southern California. In fact, shortfalls in one region act as stimuli to others.

We learn from this African case a simple and obvious lesson which seems to have been lost in our increasing tendency to seek complex rather than simple or obvious explanations. No matter how potentially productive a nation, if it is at war within itself, it cannot have a burgeoning agriculture, nor fit into any wider agricultural system, such as a continental system.

Toynbee seems to be one of the few important modern writers to stress this simple point. He offers a poignant passage on the Vikings. Their depredations along the coast of Europe ceased when they became Christianized. They became farmers. Toynbee notes the apple orchards of Brittany were a Viking creation, a dramatic change for such men. To tend an apple tree is to give up war as one's personal profession. To have a burgeoning agriculture is to give up the active practice of war. The pastimes of war and agriculture do not mix well.

To foster a free flow of trade regionally and to go beyond transportation systems and the absence of trade barriers requires also a fiscal and commercial system, integrated on a continental scale. Today, especially in the United States, modern banking systems and agriculture are copartners. Virtually no agricultural system beyond simple subsistence agriculture can exist without banking support. Such support extends from straightforward lending for investment to simple credit for personal and other operations. City dwellers should remember that farmers (and especially cattlemen) receive their incomes sporadically and must endure long periods without income.

U.S. banking in support of agriculture is a vast operation, calling for an integrated system of banking. It is difficult for a small local bank to make large loans or many loans. When such a bank becomes integrated with other banks, such as through the U.S. branch banking, the lending capability of the local bank is increased. Nevertheless the localism of a bank is essential. It knows the farmer, it knows local conditions; it can make better forecasts than can some distant partner. This intimacy is critical. When it is combined with the lending capability of an integrated system, the result is an extremely viable funding operation. One obtains the best of two worlds — the financial power of a continental or regional system and intimate knowledge born of local association.

Throughout most of the Third World, there is virtually no rural banking system. It does not exist either locally, regionally or as a continental system. Money lenders tend to be the substitute. Today, however, with computers and highspeed electronic communication, a banking system geared to great numbers of peasants becomes an extremely viable possibility. Creation of such a system is vital. Without rural credit, there will be no large scale agricultural advancement.

The size of continents such as Africa and South America in particular calls for further development of country-to-country banking and other cooperation. Such an advance is obviously dependent upon reasonable relations between countries. The only alternative is to do it all oneself. But the possibilities of country to country banking in most continents are remote because man-made political boundaries are so illogical. Nevertheless, despite the obvious need, there seems to be no progress toward the kind of cooperation necessary to create a banking system vital to agricultural development. Indeed, in regions where cooperation might achieve great results, one sees hostility, tension, and an absence of even the most elementary forms of cooperation. Neither do these countries favor outside help in financial and commonweal matters. Their state of mind, often shaped by a real or imagined colonial experience, rejects such an intrinsic enterprise as a foreign-owned or -operated bank. So moneylenders, paradoxically often foreigners beyond most controls, hold sway. The issue under discussion is simply that of internal stability and reasonable relations with one's neighbors. Banking, for agriculture *per se* needs such stability, otherwise its stimulus is lost and ongoing support for development is negated.

Nor is direct outside investment in agriculture welcomed in most Third World countries. There are a few exceptions, Nigeria and Brazil being most notable. Generally, outside investment in agriculture is regarded in terms of plantation days where entrepreneurs extracted wealth for export to a home country. But foreign capital has no desire to enter a country where internal stability is shaky or where international relations are strained. So we do not really know to what degree foreign capital might have played a role in agricultural development. Inhibitions from the colonial past and investor reluctance to take risks in unstable situations have taken care of that.

Thus capital infusion tends to become a government-to-government or government-to-international-agency matter. As we shall observe later, these relationships are not always productive, carrying within them their own inhibitors. Thus development lags, and with it also lags the growth of internal stability and self assurance. Without these, potential outside support of development cannot be obtained. Thus we have a tragic paradox, a vicious circle.

Only a government which has internal self-assurance can make policies which might in themselves stimulate development. In the United States, for example, there is no doubt that past tax laws have accelerated agricultural growth and efficiency. Tax deductions based upon investment credit, depreciation and assigning of funds for write-off have greatly stimulated development and use of new agricultural machines. Tax leverages inherent in cattle ownership have helped create the cheapest beef in the world, expressed as a percentage of take-home pay. U.S. beef, at that price, is unequalled in quality. Many expenses that would normally be capitalized in a nonagricultural business can be listed as expenses in agriculture in the U.S. Thus the farmer can have the advantage of either creating large expenses in a given year or, if more advantageous, of capitalizing these expenses and spreading them over a period of years.

Without becoming too technical, U.S. tax laws also allowed agriculture to use the cash receipt method of reporting rather than the more normal accrual method. This means that the wise farmer can move income and expense into the tax period that would be the most advantageous to him from the standpoint of taxable income. This in turn has allowed him better planning not merely for personal use of his rewards, but reinvestment in his farm. All assets in an agricultural venture, except bare land, can be capitalized and depreciated. Along with this, the farmer has the option of taking a double declining balance or a straight-line depreciation, whichever is more advantageous to him. This encourages him to produce and reinvest. Of great significance in most agricultural ventures, income and expenses can be reported in a combined statement. This allows the farmer to optimize his cash flow and his expenses, to optimize depreciables, or to transfer the taxable profit from one agricultural venture into development of another agricultural situation, and over all reduce taxable income.

The purpose of these tax laws was to stimulate agricultural productivity, and they did. They opened up revenues for the creativity of the American farmer. It is his creativity which is the very basis of U.S. agricultural developement. Any situation which curbs creativity will, in turn, curb productivity. Further, U.S. tax laws helped the American farmer share the profits in what he produced.

Third World peasants face the same situation as does the American farmer; each must obtain a fair reward if his creativity is to be stimulated. Agriculture is a tricky business. Risks are high and often flow from phenomena over which one has no control. Profits are low and uncertain. A 5 to 6% return on investment is considered good. Four percent is nearer average and some years there is no return at all.

To the farmer there are advantages, values. He sees the farm as an avenue for his personal expression through his work. He associates favorably with all elements in the symbiosis. When he talks, he talks to other farmers. They understand. The farm is his life and his home. He will suffer many depredations of nature and market, yet hang on. Then there is that remarkable person the farm wife. Thai peasants used to have an aphorism for the peasant wife. "She," they would say, "is the hind legs of the elephant." It is the hind legs of the elephant that give stability to the more fancy aspects of front-quarter maneuvering. It is the hind legs that give the ultimate push that takes a task over the top.

It is also elemental and basic that this creativity inherent to successful agriculture can arise only when a government has the self-assurance to let its people go. Sadly, successive governments are losing that assurance. Governments involved in internal dissension or international tension are also reluctant to give their farmers freedom. Their attentions are elsewhere; their fears are manifold. Thus for agricultural development to proceed, first must come national self-assurance, and this is born of governments that do not promote conflicts within or without.

Here one is forced to make another aside regarding U.S. government and farming. This government, which gave so much incentive to U.S. farmers through tax laws, seems intent upon destroying that same farmer through death duties, inheritance taxes and revised tax laws that kill incentive. Here it is the smaller farmer, inevitably involved in a family farm operation, who suffers most. As death comes to all, one can foresee the demise of the small family farm in the U.S. within the century, through government action.

Elementary internal and external stability is a basic requirement for productivity. Dissension of any kind is expensive to agriculture. History suggests that man is inclined to war and internal turmoil. This now seems to be especially true relative to Third World countries — that segment of the planet which needs agricultural development the most. Let us look in more detail at the dynamics of internal and external disturbances in a region that promises much but has delivered little.

Rivers are often a symbol. Sometimes they divide, offering "natural frontiers." This is a military term dating from the French Revolution when the Rhine was thought of as one of the "natural frontiers" of a new France. When the Japanese invaded China in the 1930's, each time in their southward movement when they came to one of the great rivers, the Huang Ho or Yangtze Kiang, they thought they might stop there, consolidate the rear and thus use the river again as a "natural frontier."

But rivers also join. They carved their way through the earth's crust long before man created national boundaries which the great rivers ignore. Since early Neolithic times people have used rivers for communication, for this joining.

Great rivers often symbolize fertility. The Nile with its annual additions of nutrients to the flood plains of Egypt was one example, almost a constant through recorded human history. Rivers have provided irrigation for crops. We now know that irrigated farming is the best kind of farming and that the future of mankind's food and fiber production is intrinsically linked with the distribution and application of water under man's control. With advent of an engineering ability to build great dams, not only do rivers offer more water for irrigation, but also sources of electrical energy.

The Mekong is such a river. The fifth largest river in the world, it rises on the edge of the great Tibetan plateau in China. In its early traverse through Chinghai Province, only one mountain ridge separates the

Mekong from the Yangtze Kiang. This latter river bends eastward and helps form one of the great eastern Chinese rice plains. The Mekong continues southward through China. It forms the border between Burma and Laos. Later it forms part of the border between Laos and Thailand. The Mekong then flows through Cambodia and into the South China Sea through "nine dragons' mouths," forming the Mekong delta in what was South Vietnam.

It is a magnificent river. In its upper regions between Laos and Burma it rolls darkly between jungle-clad mountains, cleaving narrow valleys wherein a few small villages are nestled. Between Laos and Thailand the river resembles the Mississippi as indeed does the entire Mekong basin. Near the Laotian town of Pakse there are two rapids which effectively deny large ship navigation in the upper reaches. French colonists had in earlier times thought of the river as a water route to China. The Kemmerat and Khone rapids effectively destroyed that concept. Nevertheless, ships of up to fourteen feet draft can sail upriver as far as Kratie in Cambodia, a distance of about 300 miles from the ocean.

The river provides a port for Phnom Penh, the capital of Cambodia. Near Phnom Penh a natural channel cuts westward from the river and enters the great lake of Tonle Sap in Cambodia. When the monsoon floods the river, waters rich in nutrients flow westward into Tonle Sap. When the flood subsides, the flow in the channel reverses, partially draining the lake, which when full is about one hundred miles long and thirty or forty miles wide. With the lake partially drained, there remain great warm nutrient-laden pools. These provide nurseries for many species of fish. Then the lake fills again, and with new nutrient-laden water from the Mekong, its fish productivity becomes enormous. It is said that the lake produces more protein per acre than any other place on earth.

Throughout the length of this river, some 2,800 miles in all, there is not one bridge or dam. Peasants reap a small harvest of fish from the river proper. Sometimes a few vegetables are grown in the rich silt along the banks when the river is low. There is virtually no irrigation water drawn from it. Yet, along its traverse the rice farmers are dependent upon the monsoon rain for one rice crop a year.

The rice cycle in some areas adjacent to the river is in fact quite a desperate venture. The large region known as the northeast plateau of Thailand illustrates the point. Here soil has been leached badly by rains falling heavily over the centuries. The rains however, come late and are fairly concentrated. This, along with the lie of the natural watershed of the region, creates a massive six-week flooding period. The northeast is plagued by too much water for a very short period and not enough the rest of the year.

To counteract this situation many rice farmers grow "floating rice." The water becomes too deep for ordinary rice. Floating rice is a species that grows quickly in water up to five inches per day. Thus it keeps pace with the rising flood. It has to, for the water drains away as quickly as it rose. Soon dry, hot weather ensues and the ears of rice ripen. From this method of cultivation, the farmer hopes to obtain enough rice to feed his family for a year.

When this short crop cycle of less than three months is over, a long dry season begins. The *padis*, once aflood, become so dry that it is difficult to break ground with a pickaxe. Nothing grows. The farmer waits. Of all rice cultures, the "floating rice" process is the most tenuous if not the most desperate. As one talks to farmers one senses the apprehension they have from dependence upon a process that has so little assurance to it. Yet alongside these apprehensive men, the fifth largest river in the world quietly and undisturbedly rolls down to the sea.

Many people, especially French colonists, had visions of developing the river. In terms of irrigation and energy production, the estimated potential was enormous for the entire lower Mekong Basin. Instead of one rice crop a year, two or three were envisaged. Of equal importance, areas which could be farmed only under monsoon conditions, because of terrain configurations, might also be cultivated with many varieties of crops. In addition to the irrigation project, there would be energy generation for an area virtually without electricity. Men also dreamed that with the advent of dams and locks the two great rapids could be gentled with deep water and the river could be made navigable.

In 1957 under auspices of the United Nations, a Mekong Committee was formed, headquartered in Bangkok. In 1957 the Committee included representatives from Laos, Cambodia, Thailand and the then-South Vietnam. Also involved were other persons with various specialized knowledge, drawn from foreign countries.

By 1970 the Committee had spent $100 million on various studies. These funds came from twenty-six countries, fifteen international organizations and four foundations, together with small amounts from a large number of business organizations. About one-third of the funds originated in the U.S. The money was spent to foster a number of studies and to support the Committee. Dam sites were plotted not only on the Mekong proper, but in tributary rivers in the lower basin. Flow rates were charted and, of especial importance to the lifeways of the river's nearest inhabitants, were related to potential agricultural development. In all about 40 million people would be directly affected, with about another 60 million receiving indirect, but no less substantial, benefit.

To illustrate the scale of the project, the first dam envisaged for the Mekong proper, the Pa Mong, would create a lake co-joining Laos and northeastern Thailand, about one-quarter the size of Lake Erie. At its first stage the dam was designed to generate 2.8 million kilowatts of electricity and, at full capacity, two and one half times as much electricity as Aswan in Egypt. This accretion was to occur in a region that had no electrical power. The new lake more than one thousand miles upstream from the river mouth, would provide navigation facilities, flood control and, most important of all, sufficient irrigation water to farm approximately five million acres all year round.

This enormous facility was but the first dam envisaged. No clear plan emerged as to others, but at least three more of like size or larger could be contemplated. There were also subsidiary dams envisaged on tributaries, and these promised to be at least as important as the main dams when assessed in a regional context. Nevertheless, great dams do not make farms flourish on their own. Only farmers can do that. Neither does it seem possible that the region could have utilized fully in the short term the enormous increment of electrical energy that might ensue. Thailand, for example, the most electrified country in the region, consumed only 680,000 kilowatts in 1970. These kinds of problems are ideal problems to confront mankind. How to utilize resources, rather than how to survive without resources?

The Mekong project has gone nowhere. Four tributatary dams have been built on national territories. Two more may be constructed on national territories. The Mekong rolls on to the ocean, as pristine as of old. The monsoon still dominates agriculture, and "floating rice" farmers still ply their desperate venture. They hope that the floods will not rise or go down too fast and condemn their plants to extinction. Each year they hope their luck will hold. Neither has electricity come. Small villages along the river in Laos, in Cambodia, Thailand, and Vietnam are still faintly lit by oil lamps. Everything is as before. Only the potential continues.

Reasons for failure of the Mekong project are as obvious as they are distressing. From the beginning, even within the Mekong Committee itself, tensions between the involved riparian countries inhibited forward movement. Members fought and bickered over old animosities usually quite unrelated to the development of the river. Animosities between Thailand and Laos, Cambodia and Thailand, Thailand and Vietnam, go back for centuries. The Cambodians still regard the Thai as vandals who sacked Angkor in 1456, this being the last of numerous Thai attacks. In much the same way, Cambodians regard Vietnamese as invaders who have in the past taken their land and taken Cambodians into bondage. The Thai, in particular, and the Vietnamese, to a lesser degree, return the Cambodian sentiment, with Cambodia as villain. The Lao regard all three as predators in the recent past.

During the early period of the Committee's work, Cambodia did not have formal diplomatic relations with Thailand and Vietnam. Regrettably, in recent years the situation has worsened, this deterioration being brought about as a result of the Vietnam War. Cambodia has now closed its borders to virtually everybody, returning to an xenophobic state that has frequently beset that country. Reports indicate that through the use of force there has been a massive relocation of people and genocide on a massive scale. The idea seems to be to disperse people from town to country in what will certainly be a futile attempt to rehabilitate and develop Cambodian agriculture. There also seem to be strong ideological overtones to this movement. The dogma is that cities, there is really only one in Cambodia, are corrupting, but rural life and work on the land are cleansing and uplifting. Cities breed individual greed and avarice. On the land one gives of oneself for the good of all.

Movements, such as seems to be underway in Cambodia, especially under coercion, have proved especially unsuccessful in the past. More is involved in successful agriculture than the curbing of greed and avarice, which is not necessarily the prerogative only of city folk. In successful agriculture, individual initiative is the key and its rewards are not only material. Throughout history every farmer worthy of the name does attain some peace and satisfaction with himself in successful husbandry. He receives stimulation of mind by the prior planning that leads to growth and fruition. He achieves equanimity by cooperating with nature, rather than being forced, coerced, to face nature on someone else's terms. Today Cambodia is seemingly turning inward, uninterested in harnessing the riverine bounty that rolls by Phnom Penh's back door, because this has to be done in cooperation with others. Who can deny that the Mekong development scheme would have been a more fruitful way of benefitting Cambodian agriculture than rhetoric supported by cruel, destructive punitive acts?

At this time Laos and Vietnam are paralleling the Cambodian internal policy, though perhaps in less drastic terms. Perhaps it is all only a catharsis, a recovery symptom as they emerge from the long and disruptive Vietnam conflict. It is difficult to imagine that Laotians, with their traditional emphasis on individual autonomy, will respond productively under a rigid, inward-looking discipline which seems to have little purpose in terms of the daily well-being of ordinary persons. We shall have to wait.

Thailand has also had internal problems. Another attempt at a democratic system has failed and the country has returned to its old pattern of military rule. Or has it? An answer to this question has enormous significance to several million Thai rice farmers. Anyone who has witnessed the tension inherent in filling the family rice basket for one year born of floating rice cultivation has to exclaim, "We can do better than this!" Again we must face the ghastly question. Will the floating rice farmer have to wait while city-based leaders work out their own political peccadillos? Probably! Almost certainly! Meanwhile the Mekong rolls along.

Looming over the internal situation of all these states are their external relations with one another. And hanging over them all is the People's Republic of China. These small Southeast Asia states do not fear a Chinese army marching across their borders, but the non-military impact that a strong China has always had upon small Southeast Asian states. Each of these states has large Chinese minorities, the Overseas Chinese, and all have restrictive immigration laws preventing an influx of Chinese. Will there be pressures from China to relax these laws? This would be one question. Where will long term loyalties of this Chinese minority be directed? Today, the Republic of China on Taiwan holds a favorable position in the minds of many prominent overseas Chinese leaders. What if Taiwan ceases to be an autonomous country?

How will the People's Republic of China react to the internal conditions of each state? Will the new military regime in Thailand have to face a guerrilla movement supported by China? It clearly faces such a movement supported in terms of training and equipment by Vietnam and Cambodia. These fears of many Thais may not be realized. China and Vietnam may be too preoccupied by their own affairs for active pursuit of political and military ends in far-off places. This will probably be the case — but the probable result does not really matter today. It is fear that counts. It is apprehension that something adverse might happen. It is unkown disaster that inhibits.

Beyond building various structures to harness the river, to obtain maximum benefits other non-material factors would be needed also. The entire region, regardless of national boundaries, would need to become a coordinated sub-continental agricultural system. This calls for free flow of trade and commerce. Transportation must join freely. Energy outputs must flow freely. Tariffs should be reduced or disappear as should most petty customs restrictions. Above all, persons should be able to move freely about the entire region on their daily business. Only then would the great bounty made possible by the dams be realized.

Seasons will come and go. The Mekong River traverses a subcontinent. Great monsoon floods turn the river into Eliot's "great brown god."[1] There will be a dry season when sand banks and islands dot the water like warts on a wrinkled skin. Peasants in contiguous lands will live and work as they live and work now. Their imaginative minds will never receive that stimulus that the river could bring. The results of their stimulated minds will not spill out in productivity as yet unimagined.[2]

To date, the Mekong River development project embodies a relatively well-defined form of ancient human tragedy. We can see the Mekong syndrome being repeated elsewhere. This is especially true of the Middle East, Africa, South America and the Indian sub-continent. Each scene of internal instability and external dissension has its own dynamic and its own constraints. Historically we find these instabilities to be endemic to the human condition everywhere. We simply do not have data to measure how much human effort and treasure has been expended upon it all.

Neither is the point being made that internal and external instabilities are the natural lot of man. Rather, the point is that man must surmount these more barbaric of his pastimes, and in spots has done so. What can be done once can be done again. While mankind shows a great proclivity for turmoil within and without and pays a terrible price as a result, man also can stabilize situations and as one result have his fields burgeon.

One explanation for why we do so badly in much of our agriculture is our inability to stabilize. This is but one reason, but there is so much historical and current evidence of it that we are forced to note internal struggle and international tension as an inhibitor of good husbandry. Not only that, this relationship and man's knowledge of that relationship dates into antiquity. Biblical admonitions suggest the antiquity of man's harm to agriculture by his wars and tensions:

> **They shall beat their swords into ploughshares and their spears into pruning hooks; nation shall not lift up sword against nation, neither shall they learn war anymore.**

Isaiah was pointing out through prophecy that until swords are transformed into articles of husbandry, people will be deprived of bread.

What is to be done? We know what to do — we should stop this internal and external squabbling, or if we cannot, at least we must tell the truth about it. Instead of some spurious reason as to why we do not feed ourselves, we must admit that here is at least one reason for the shortfall. To look at it any other way, to blame over-population, weather, poor soils, or such like is to take one's eye off the ball. If mankind then fails to recognize the real solutions for the dilemma, he can hardly expect to find a solution.

The issue in this chapter is not agriculture *per se*, but what international and internal tensions do to it. In most instances, agriculture is victim of the malaise. What then is to be done, for we know that man does not have to live in a state of tension, one with another? It bears repeating: we must speak the truth. We might also go further. We need to propose imaginative schemes to improve man's material condition. The Mekong River scheme was one example, yet few people know of it as a proposal and fewer still know why the scheme came to naught. It is an imaginative scheme coupled with truthfulness that offers us a possible way out of this unfortunate dilemma. Let us look at yet another scheme — one which is certainly imaginative — and put ourselves in a position to portray its potential to everyone.

This scheme involves a most extraordinary concept, put forward by a Japanese. The Persian Gulf is a large, but fairly shallow, sea. Around the tiny Gulf cluster such countries as Iran, Iraq, Kuwait, Saudi Arabia and the tiny sheikdoms of Bahrain, Qatar, and the Trucial Omans. The proposal is to build a dam near entrance of the Gulf. The Gulf would then be drained of seawater. Anticipated oil resources under the Gulf would be tapped and piped to shore. The Gulf would then be allowed to fill with fresh water from the Euphrates River, the lower reaches of which would be deepened and diverted.

This large accumulation of fresh water accumulating in the Gulf would dramatically transform a vast littoral which is currently arid and barren. Through irrigation one could expect a result comparable to those in the American southwest, a 365-day growing season and high yields per unit

The Japanese scheme may be little more than a dream, but it does illustrate two points. First is the need for that difficult and elusive quality, an imaginative mind. Second is the absence among us of imaginative minds. This is especially true among many political leaders, who seem to have no imagination at all. Perhaps it is the nature of their role. They too often see only the immediate in terms that further their immediate interests. There is not among them even a Louis Napoleon, a man whom contemporary history tends to despise, but who at least accepted a conception which led to the digging of a great canal.

Oddly enough, imaginative design might further the ends of those same politicians who so avidly seek to obtain and hold power. People everywhere today seem to yearn for a noble goal, a real tangible goal as opposed to rhetorical utopias promised so often and never acquired.

Yet who among us would say that genius no longer resides in our midst, that genius flowered only in the past and will not come again? It still exists, but it appears to be stultified because it is; stultified by erosion of individual freedom, without which activation of genius is impossible.

Thus we have analyzed the first inhibitor of agricultural development, that is, internal instabilities and international tensions. The latter is the important one at this stage. It is petty squabbles between small states, the roots of which are often lost in antiquity, that inhibit their agriculture most. These directly affect the capacity of persons to make use of the planet's agricultural potentials, and of humans as individuals to reap the reward. This is the real price...this deleterious impact upon human capacity and need. Without a human dynamic nothing functions. Internal instability and international tension can greatly inhibit that dynamic. It is but one inhibitor, and not the most important. Let us look at others, some of which have a greater effect.

FOOTNOTES

1 T.S. Eliot, *Four Quarters,* Dry Salvages, 1. (New York, Harcourt, Brace & World, 1963).

2 In April 1978 I visited new irrigation projects in Thailand. Two of the dams built on the Korab Plateau were tied in with the unborn Mekong Scheme. The effects of these water works was electrifying. Farm incomes had increased 800 to 900%. The most striking effect, however, was that the reason for this was not just water in an arid region. In striking fashion the Thai farmer was basking in the ethos. He was calling the shots. His government was listening and serving. If Thailand can spread irrigation and the ethos, it is due to follow Taiwan into economic take-off.

AN INHIBITOR: IDEOLOGY 7

Ideology and farming do not mix very well. To paraphrase Burke, "great ideologies and little ventures go ill together." For agriculture to be successful, it has to be parochial and intimate in its interests. It must be flexible because elements in the symbiosis take on different values in different situations. Where it is allowed, it is also dynamic. Even in its dynamism, however, there is a constant empiricism. The only absolute is that we recognize the correlations of different features of the symbiosis.

Ideologies do not fit well in such a framework, but men are prone to ideological pursuits. Often on capturing an ideology, they turn it into a force compelling other persons in a particular way. Ideology as a force frequently constrains people and attempts to divert them toward paths they had no intention of treading. This is a feature of our century... men who want to direct because they have found "the way."

In this book ideology is given a particular meaning. The concern is with political ideology, an ideology embraced as the root and motivator of a political system. This leads to consideration of political ideology in action, where the ideological construct forms the basis of that action. Thus we are concerned with those political ideologies that tend to be tautological. Here alternate premises are not admitted. Thus the basic construct does not change, for it cannot. It takes upon itself the mantle of immutable truth.

Not surprisingly, such ideologies tend to be totalitarian and dictatorial. In line with Hegel, they reject contrary empirical and scientific evidence as they do traditional logic. Those who do not embrace ideologies are wrong and are kept outside the political system. Persons who do embrace this pattern of thought seem to need a rigid psychological set which will abolish all doubt and provide pre-designed answers to all questions. They demand that all other persons adopt a similar mental set. The closest historical precedent for this situation is when religion lapses into dogma and where a revealed truth predisposes its owners towards zealous reform of societies, even by fire and sword.

But the ideological constant referred to here does not resemble religious idealism in its tautological sense. The synthesis of the Christian faith (and beyond belief in the Trinity) is personal acceptance of the divinity of Christ as Son of God. This is an autonomous act of the individual, not forced externally by another human. Christianity also allows persons to return to disbelief, again autonomously, and autonomously to bear the penalty of judgment; not judgment of man, but of the Divine.

Buddhism asserts more positively that it is the autonomous individual who must make free choice in terms of his endeavors to follow the Noble Path. Not only that, but it is the individual who must work out his own salvation through individual action derived from his own thought process. The Buddha was not an intermediary working on behalf of an individual. He merely showed a Way. Followance was an individual decision. Reward or penalty was directly and intimately related to the diligence and skill with which the individual did or did not follow the Way.

Only in Judaism do we find a variation upon the free will of the individual to determine his own fate. In Judaism God can intermittently act in history to change man's destiny. Even then God is not absolute in dominating men's lives, for men have freedom to reject or ignore God's demonstrations. Neither does God act as a totalitarian. He may from time to time show signs to mankind and offer symbols which may enhance human understanding. But it is the right of man to accept — or to abstain. God also remains a mystery in history.

> **...history is unredeemed, why God 'hides his face,' or is, as it were temporarily without power and in any case restrains his Messiah from coming. This does not, however, either refute Jewish faith or deprive it from content, so long as the promised coming of the Messiah can still be expected.[1]**

Above all, in the Jewish faith, man is in dialogue with God, can even argue with Him as we see in the *Talmud* and in *Micah* and *Job*.

Thus religion as ideology is different from contemporary political ideologies. The political ideologies we must deal with today disregard individual idiosyncrasy and, indeed, exorcise such. In modern political ideologies, which reject individual autonomy and assert grand designs for all mankind standing as an incoherent assemblage, it is but a short step to terrifying simplicities: mass education according to a single formula, mass activation, a mass goal, mass arrest, mass deportation, and mass slaughter. In the end, political ideologists in their disregard for the individual become in Orwell's words, "men who think in slogans and talk in bullets."

Basically, the Third World has no ideology of its own. It has listened to and, in some instances, has embraced versions of European ideology or of Maoism. The principal one has been Marxism, and to a lesser extent its younger brother, Fabian socialism. It was not difficult for these ideologies to implant themselves, no matter how imperfectly. They really had no rivals. There was no indigenous system other than two religions, these being firmly held in only very few countries. Islam in the Middle East in the main resisted. But can the Middle East with its mounting riches really be seen as being in the impoverished Third World? To a much lesser degree Buddhism, both Therevada and Mahayana, has in East Asia and in Southeast Asia offered an alternative way of thinking.

Its resistance to European Marxism has not been vigorous. Lack of militancy in Buddhism, Vietnam excepted, makes it a poor counterbalance to determined Marxism. Even attacks on Confucius by the late Mao Tse-tung were less an attack on a viable, operational doctrine than they were on a legacy much of which either pre-dated Mao or simply refuted Mao as an oracle from whom all past and present wisdom flowed.

In particular, Confucianist stress upon development of the individual, albeit within the family, was a primary barrier to the elitist totalitarianism of Mao, and the imposition, through dictatorship, of collectivist rather than individual values.

Thus Marxism as an ideology has flowed into the Third World almost by default. It had few counterbalancing constructs to oppose, refute, or offer alternatives. Further, Marxism offered something to people beyond promise of material reward. Nevertheless, peasants in particular have had a high degree of skepticism regarding Marxist promises of material enhancement. They have heard so much of this before, and through paucity of result it has always passed like a summer shower, leaving little trace. But Marxism did offer something more important. It offered a construct, a plan within which men could see themselves. It gave the individual a sense of his identity, as a living and breathing actor in an historical plan. Man could identify himself in time, see where he had been and where he might be going. The urban Western mind may balk at this. Not peasants, the Western mind would say, not these dumb, illiterate, unlettered folk, concerned only about eking out a precarious livelihood. No such sophistication could possibly be held by peasants!

To the Western thinker, Marxism's only appeal to peasants was material. Prospects of a full rice bowl and more than that for their children — this was the appeal. Urban Western man so firmly believed this that one of his countervailing responses was to offer to these unfortunate peasants the option of a fuller rice bowl than they might obtain from Marxism. Urban Western man is still puzzled that the peasant did not respond to the promise of greater material largesse. In fact, material largesse had little, if anything, to do with it. Let us illustrate the point, this sophistication of peasants as regards ideas; and in so doing, perhaps illustrate the lack of sophistication of urban Western man.

Angkor in Cambodia represents one of the world's largest man-made ruins. Its buildings are scattered over 200 square miles, among thousands of forest acres. All of the buildings of Angkor have a common form. They are laid out to conform to a preordained plan and to meet a purpose that could only be achieved through such a plan.

The construction of Angkor began in the 5th Century A.D. and the area was finally abandoned during the 15th Century A.D. The great stone buildings remain. The wooden construction, housing ordinary people — merchants, food peddlers, carpenters, stone masons, fishermen and farmers — have long since rotted away. Two great artificial lakes remain. These were the key to the astonishing irrigated agriculture of

Angkor, with crop cycles that never ceased.

As noted, main buildings were laid out according to a cosmological plan. To these ancients, as is the case with many today, the firmament, the stars in their places, told man where he was and where he was going. It was a serious business, this astrology. Crops were planted and harvested in accordance with it. People birthed, married and died by the stars, and the stars directed their lives in between. Games of contest were set by the stars. Prospects in war and diplomacy were so guided. The Divine King was dominated by cosmology in every waking hour from birth to death.

In large measure architectural schemata of Angkor as a whole, and that of each individual building, is an earthly translation of the cosmos. It is the cosmos brought down to earth, in an enduring and understandable form. The great temple of Angkor Wat, the single largest edifice, still stands. The great moat that surrounds the edifice is the Sea of Churning Milk, which cosmologically represents the boundaries of the Universe. The universe is then portrayed in stone in the confines within. At lower levels and in outer reaches of the building, we see common things, bas-reliefs of man himself. People fishing, cooking, children playing, male and female courting, and warriors marching and clashing. As one moves toward the epicenter and climbs upward in the temple, stairs become steeper, just as progress through life itself becomes more difficult as we pass through it. We mount to five great towers, each with a name and a symbolic meaning. These surround the celestial center of the Universe, another larger tower symbolizing Mount Meru. High on Meru, at the real epicenter, are the mandala, magic circles each with its intrinsic message symbolizing the Void.

This cosmological map was designed on earth so that men, ordinary men as well as priests and soothsayers, could locate themselves in time and space. A man living in Angkor in, say, the 11th Century, could identify himself and where he was in the pattern of things. If he was adventurous, he could also see where he might be going. He could see his historical past. It was like having a road map which provided assurance by showing where one was and what might lie ahead.

Today we can see that for the ancient people of Angkor, their astrology and cosmology — the ideology by which they lived, and which was symbolized in the monuments they built and left behind them — all this was fundamentally illusion and self-deception. For at the same time as their priests and soothsayers were guiding every major event in people's lives by stars and planets in the skies above them, the very economic basis of their whole civilization was, in truth, rotting away under their feet. For the collapse and eventual disappearance of the Angkor people and culture was most probably caused by failure of their agriculture, due to soil degeneration from manmade causes.

Thus, that absolute essential to human life and therefore to human culture and civilization, food to nourish their bodies, was denied to them in the end. The total irrelevancy of an ideology, a religion, if you will, or at the very least a cult of belief, to the realities of basic agriculture, could not possibly be better illustrated than by the Angkor case. In the end, the ideology was helpless to show where the Angkor people were going. Its prophecies were hollow falsities, conjured up in the minds of the learned priesthood which had tenure of the ideology under which the Angkor governmental system was operated.

All this, we must remember, was a thousand years ago. In the meantime, the Angkor people and their civilization vanished completely from the face of the earth, victims of a total ideology which had no ultimate power to save them from their bitter fate. Their ideological belief-system simply led them to their doom.

Marxism, like most strong ideological constructs, is a little like Angkor, but in a political sense. Marxism gives man a view of history, where he has been and what it was like. Marxism tells man where he is today. More importantly, Marxism tells man where he is going. It gives to an individual shape, form and purpose, even proof of his place, and therefore his personal significance in history.

We must assume that the average man in the street did not see his place in Angkor with such clarity. He was probably more interested in the games. Men today do not see themselves in their current historical context, either, but Marxism helps as Angkor helped. Some of the cosmological road map must have rubbed off upon humble people of the time. This is a reasonable assumption, because people do not like being lost. They like to know where they are and where they are going. In what urbanized Western man calls primitive countries this need is not less; it may even be heightened. This may be the main appeal of Marxism. Not only did it have few philosophical or doctrinal rivals in modern times, but it also fulfilled a psychic need. It gave often rootless people a sense of their place and their purpose.

It is difficult for the urban Western mind to accept the fact that this notion, subtle and esoteric, applies to humble peasants. Peasants have imagination just as vivid as do the educated, and sometimes even more so. Peasants have psychic needs too, no less than those of the seemingly more sophisticated. It is to this that Marxism sometimes, perhaps crudely and in a non-erudite fashion, addressed itself. Any other ideology as didactically constructed might have done equally as well. The pragmatic, empirical improvising man in the U.S. really had no counterforce to offer. Most Americans, for example, are neither ideological nor anti-ideological. They are a-ideological. Especially so in terms of the agricultural symbiosis. It will be suggested that this is a great strength. It should not, however, blind us to the impact of ideology upon people we consider backward.

Thus, ideology is abroad. It is abroad in the world at large, and in the Third World. It is not merely the province of the intellectual. Its impact, however, is at the same time more intellectual and more subtly psychic than it is material. Here, then, is a general setting, as seen particularly with Marxism in the Third World, for probing ideology a little further before relating it to the agricultural symbiosis.

We must also, in our American a-ideological way, give a construct to ideology not only as a theory, but as an act. Let us briefly look at the theory as it affects men.

We have noted that men who embrace ideology as a dominant factor in their lives seem to have a moral and psychological need to frame a construct for human affairs. Within this construct the premises are immutable. They represent a given, but total, truth. Being truth, the premises cannot change, and to suggest any such change is heresy. Heresy in turn must be exorcised because it endangers a total truth. The ideology becomes a closed system admitting nothing new, for why should it? At the same time ideology demands as its right the role of arbiter of human action as well as of human thought.

We see the most striking example of this concept of ideology as a closed system in Hegel, who in turn had such a marked impact upon Marx. Hegel enunciated an ontology, an abstract essence of being which in turn admitted no refutation based upon such scorned notions as new, practical evidence. A scientifically proven fact which might dispel the ontology was not to be admitted. For Hegel such childish things as empirical data were similarly ignored.

Ideological reasoning is the reasoning of men who have discovered truth through intellectual reasoning, but not truth born of practical experience. Once having grasped truth through intellectual means, the ideologue shows his most pronounced characteristic: he must use the truth to shape and mold other humans, even if they be humble farmers. The urge to mold men takes many forms. It can range from the manipulation of didactical evidence to that of making great, swooping, voracious raids on the lifeways of millions of people.

Let us look at the intellectual distortions that political ideology can demand, in this instance one connected with agriculture. Science might indicate that heredity seems unaffected by environmental changes in preceding generations or that, if it is, change comes slowly and often with unpredictable results. Mendelean genetics is based on the concept that in plants, as in other living organisms, there is obedience to simple statistical laws about physical facts. The principle is that in the reproductive cells of hybrids (Mendel's model) half transmit one parental unit and half transmit the other.

From this simple statement, one then advances to biological complexity. When there are several pairs of alternate characteristics, these enter into all possible statistical combinations in the progeny. There is, however, a law of independent assortment where up to seven pairs of differentiating characteristics combine at random. These are variations proven again and again in experiments. Independent assortment applies only to genes that are transmitted in different "linkage groups" as chromosomes into which genes are organized. Also the appearance of dominances in the hybrid offer other statistically recurring characteristics. While the notion did not refute evolution, but rather confirmed it, it also suggested that evolution proceeds systematically and in accordance with preordained laws.

In the Soviet Union, especially between 1948 and 1964, T.D. Lysenko disputed this basic concept of biology. The basis of the Lysenko theory, which had significant impact on Soviet agriculture, was that heredity is the result of environmental changes that have been assimilated during the course of preceding generations. The idea of natural selection inherent in the older biology, and now supported by new Western research into gene biochemistry, was discarded. Indeed this form of biological theory, based as it was on the chromosome, could not even be taught in the Soviet Union between 1948 and 1964.

As offshoots of this theory, Lysenko claimed that new types of hereditary development cycles in plants could be obtained simply by regulating the quality of their nutrients. This obviously called for a dramatic shattering of the innate conservatism of every plant so that (in theory) it became totally subservient to its environment. Even by 1948 the Soviets, or those of the Lysenko school, were heralding a world-wide revolution in biology and therefore in agriculture. Lysenko's primary refuter, N.I. Vavilov, had already been arrested and exiled to Siberia, where he died in 1943.

But nature had its revenge. In practical form it refuted Lysenko and his notions. One such result, the effects of which may still be felt, was responsible for toppling Lysenko from a position of real eminence. This was Soviet failure to utilitze hybrid corn. Indeed, under Lysenko's theory, the whole notion of hybridization was irrelevant because it represented a reordering of a plant's structure in accordance with the older biological theory based on the inherent character of chromosomes.

To Lysenko and his followers the environment was the key factor. In a sense Lysenko and his political supporters, chief among whom was Stalin himself, were products of an ideological construct, and a Hegelian one at that. In essence, Hegelianism posits a unitary solution to all problems of philosophy. When Marx later translated Hegel's idealism into materialism, the notion developed of man's advancing continuously into higher stages of perfection — the actual development in fact of an Absolute. Then it was not too large a transference to assert that within the environment of an appropriately convened state, man could be conditioned by that environment to move toward perfection. The concept of the Soviet state producing the "new Soviet man" was no accident. As with plants, man could be remolded by producing the right environment for him. In more recent times, the notion that each man holds innate genetic characteristics not always affected by environment has been somewhat redeemed in the Soviet Union, and has caused pause to the construct. Also, the Soviet Union now vigorously grows hybrid wheat.

The point is that even scientific and empirical evidence does not always temper the ideologue's mind — not until disaster has arisen (as when Lysenko's theories were applied in Soviet agriculture). The ideological mind expects and demands that humans fit its construct. To persist in one's old-fashioned, perverse habits in the face of conditioning environments, represents a mutating force — something to be exorcised. As we know, exorcism in the name of ideology can become massive in its scale and ferocity.

We may therefore say that historically, ideologies, especially such well-developed total ideologies as Marxism, have treated conflicting evidence as aberrations, mere accidents to be ignored. Should the aberrations persist, however, they must be summarily eradicated where possible.

This would seem to face the ideologue with several different situations. Have accepted an ideology totally, and then to see it refuted in some way, causes the ideologue to reject the contrary evidence rather than to assume that he is at fault. In other words, the ideology takes precedence over man. Many commentators on Soviet affairs, especially Arthur Koestler in his book *Darkness at Noon*, noted this phenomenon and gives it as the reason why hitherto high party functionaries, when charged with trumped-up crimes, would confess and ask for a death sentence for crimes they had not committed, just so the ideology could be protected from their innocent, but aberrant, behavior. On the other hand, in the face of evidence contrary to the ideology, men may break. The center point goes out of their lives as the ideology is diminished or destroyed. Here is the phychological risk inherent in embracing a total idea. It may be wrong — and then what? The human psyche eschews facing such trauma.

Other reactions may also result. The holders of a firm ideological position may react quite violently (and they have) when they see, or even sense, that the ideology may be wrong. The elite ideologue immediately reforms the construct and offers to the world a new construct of his own making. But these reformers really only change words. In mind, attitude and action they remain the same. Would Trotsky have been more merciful than Stalin? Perhaps. But only in variation of method, one senses, because the construct was there, impelling him to force men to become different.

For the ideologue, the most difficult situation of all is for him to renounce ideology without substitution. To return from the psychic security of a dogma to the changing free flow of empiricism is indeed disturbing. Not only does he lose the comfort of certainty, but also perhaps a little of his self-esteem. To be a true ideological believer calls upon one to be something of an egoist, a person who believes that he is extremely important, for after all, he has found a prescription to remake the world. Certainly ideologues are rarely modest in their demands. Bakhunin, the Czarist officer turned revolutionary, friend, then antagonist of Marx, embodied this attitude well. He envisaged a tiny group of revolutionaries, wedded entirely to violence, who would destroy all institutions of religion, politics and economics, and other social organisms. This would usher in a utopia comprised of freely associated groups of humans, united in a loose federation, each doing its own "thing." But Bakhunin was an extreme, perhaps? In the words of one biographer, "The call to revolution was in his blood, as some men feel the call of sea and hills."[2] One cannot deny the terrible egos of the ideologues, their sense of rightness, and thus their inability to temper actions with mercy and never to accept criticism.

The ideologue also sees himself as a prophet. Marx certainly thought of himself in prophetic vein. He was capable, he believed, of seeing not only where man had been, but where he was to go. Not only that, Marx delineated in detail his construct of human dilemma and salvation. Thus men who follow Marx demand reward in status and position, if not intellectually, then politically. Such men do not like to be argued with o disobeyed, least of all by humble folk not given to hewing to a particula line.

There is also, one feels, a thin demarcation, if there is a demarcation at all, between ideology and dogma. Hobbes made the point,

> **The fault lieth altogether in the dogmatics, that is to say, those that are imperfectly learned, and with passion, press to have their opinions passed everywhere for truth.**

Hegel, Marx, Bakhunin, to mention a few examples, were learned. This was probably their problem. Their kind of learning inhibited them from knowing how things worked. Who knows what Marx might have done if he had fled his Highgate dungeon and learned of a world at work rather than a world as a theory? He might have modified, and pehaps he might not have written at all.

But the Hegels and the Marxes die. Then along come other men, men of implementation, perhaps as Hobbes observed "imperfectly learned" in another sense. To them the demarcation, if it existed at all between ideology and dogma, tended to disappear in a set of instructions that it must be done and done this way. To them the construct is less significant than the action arising from the construct. The action is not subject to refinement. Such is unnecessary. "Their opinion" must be "passed everywhere for truth," and by their nature, dogmatists in action tend not to tolerate disobedience.

Whatever one may call them, ideologue or dogmatist, their cult has had an extraordinary impact on significant segments of the modern world. To assume that something so widespread as agriculture could have escaped the impact of the ideologues and dogmatists is to misconstrue the history of our times. To assume that ideology will in the future play no role in the massive agricultural transformation which must occur if the planet *is* to feed itself is to assume that history has no continuum or that we learn from past mistakes.

As we proceed, it is important to relate ideology to the practical sense of doing things, in this case to farming. This is not often easy for the modern mind. We have become fragmented. To a very learned agriculturist, agriculture becomes a technical problem and often he knows only one tiny aspect of the technique. He finds it difficult to relate his experience with technique to something as seemingly alien as ideology. Yet it is related and that relationship is the substance of this chapter.

Ideology, especially when it becomes dogmatism, has been presented here in an adverse light. Many will protest and proclaim it as a progenitor, a catalyst bringing mankind into a better state. Without ideals, it will be asserted, there is little chance for mankind to advance beyond the worst aspects of his nature. It is the ideal which proclaims the humanity in man and his advancement.

Plato will be brought forth and much made of his basic concept that the ideal exists for every human construct, material or otherwise, and that there is (or should be) a sense of striving for the perfection inherent in everything. True, perfection may not be attained. Without the ideal, however, a timepiece measures only imperfectly the seconds, minutes and hours of each day, and not the time of the universe. But the ideal is within the timepiece and so remains. So it is with all things.

It will be said that without the ideal, we are nothing. We must pause again, therefore, and distinguish between an ideal and an ideology. Plato, often listed as the first Idealist, was not an idealist in the sense of the modern ideologues. Plato did not posit the forcing upon men of some absolute created by his mind. This the modern ideologue insists upon. Plato did not offer man schemata relative to which the masses must conform. Instead he held out the ideal inherent in all things as the object toward which each individual might harmonize his soul. It is the modern ideologues who have perverted Plato's ideal state into a political affair. We must note how Plato has been prostituted. Thus it will ever be when something as important as ideas and ideals are presented to humans. They will be taken and twisted into that construct or absolute to which, the ideologue insists, all men must conform. Perhaps the best safeguard is to seek truth as individual persons within individual consciousness. In the Dhammapada the Buddha gave another view of this matter which held a warning:

> **We are what we think**
> **All that we are arises with our thoughts.**
> **With our thoughts we make the world,**
> **Speak or act with an impure mind**
> **And trouble will follow you**
> **As the wheel follows the ox that draws the cart.**

This then is enough. One need not renounce ideals because they may be perverted by ideologues. Rather one should seek them for oneself.

They are, however, to be sought out by individuals and not imposed upon them.

Agriculture is a practical affair. Plants, animals, soils, climates and work of the symbiosis are doubtless a-ideological. The human work factor, however, can be perverted. It is directly and indirectly affected by ruling passions or constructs, whatever these may be. In the Soviet Union we see these passions and constructs impinging on agriculture on a scale so vast that it calls for new exercises of the imagination. Initially we must qualify any analysis we make of Soviet agriculture. We know only a little about it. Because of the Soviet treatment of statistics for propaganda purposes, accurate facts are hard to obtain and cannot be compared with those emanating from the West's agriculture. More importantly, no Western scholar has ever studied in detail a state or collective farm, especially in relation to its human dynamics. Elsewhere Westerners have lived among all manner of agriculturalists in strange and remote corners of the planet. Their research has often given us detailed, sometimes exhaustive data about practical conservation of the lifeways of all manner of people. We have nothing comparable concerning agriculture of the Soviet Union. We have statistical data usually expressed in national terms, but never in personalized terms.

On the other hand there are stories, usually sad and terrible, of the excesses of various kinds that occurred under collectivization. Even here in terms of a total comprehensive picture we face a gap, especially with regard to how the people involved went about their daily business as so many millions still do in the Soviet Union. Soviet agriculture is probably not Socialist agriculture at all. Socialist agriculture was the original intent. It was to be a system where ownership was collective, but so also was to be decisionmaking. It was to be individuals on the spot who made whatever decisions were needed to make agriculture work. Soviet agriculture, however, became a state agriculture. Decisions were made by detached bureaucrats.

Stalin in his various critiques of Soviet agriculture, and especially relative to mistakes assigned to his subordinates, always alluded to the fact that in agriculture as in all else in Soviet life, they were ushering in communism. Communism had not arrived, he reminded them, but that was the goal; that was what the process was working toward. Presumably with the advent of communism would also come Marx's famous "withering away of the state." Presumably in this condition the collective farm would be truly autonomous with all decisions and control in the hands of the cultivators.

Obviously such a situation has never eventuated. Instead in Soviet agriculture we have stateism. This manifests itself in control and direction by state planners and other bureaucrats who certainly in earlier stages acted directly at the behest of senior officials of the communist party and presumably still do. Soviet agriculture therefore can hardly be analyzed as socialism or communism, as the direct result of an ideological construct in action. The ideological purist would probably acknowledge that rather it is the result of ideology breaking down. Neverthless the root of the matter was ideology.

Historically we see five great epochal events in Soviet agriculture. First came the revolution of 1917 and its aftermath, the civil war. Agriculture was wracked and production fell by at least one half. The second shock was the shock of partial recovery. As is well known, V.I. Lenin in the early 1920's introduced the New Economic Policy (NEP). The NEP was an agricultural success. Agriculture under NEP was based on the family farm, a peasant farm. Surplus produce was marketed openly and through orthodox means of supply and demand. Peasants (as they do everywhere) showed all their traditional sensitivity to price. In a very short time, just a few years, agricultural output reached pre-revolution levels. What the future might have held for this production system is not known, for in the late 1920's and early 1930's it was destroyed.

How could this happen, this destruction, rapid and total, of a production system that had performed so well? It was a case of ideology coming into action. The creed included beliefs that contradicted the NEP agriculture. First, NEP agriculture was essentially capitalistic. It was fostered upon the pecuniary skills and acquisitive aspirations of individual peasants. It represented not an ushering in of the new, but a flowering of the old. It refuted the creed of common ownership of the means of production, distribution, and exchange. In addition, the theorists, communist leaders, had embraced wholeheartedly the concept that large, highly-mechanized farms would offer greater returns per unit of area than did ten-, twenty- and thirty-acre plots of peasants. They could not conceive that "commodity production and exchange" solely between private parties might be an efficient system. Indeed, to assume that private transaction might be more efficient as a stimulator of productive systems than a collective system was a heresy whose physical existence struck at the basis of the entire ideology. It is also probable that the hierarchy, innocent in the ways of the land, truly believed that collectivization could do no other than usher in massive increases in yield needed to help support growing Soviet industrialization.

Collectivization was instituted throughout the late 1920's and the '30's. The NEP was destroyed. The human cost was incredible in its scale and devastating in its horror. Undoubtedly millions of peasants died. The Kulaks, the wealthier peasants, were the chief sufferers. A Kulak might be thought of as a man possessing more than twenty-five or thirty acres. Winston Churchill states in his memoirs of World War II that Stalin personally told him that collectivization cost "seven or eight million Kulak lives." There were straight liquidations. There were deportations. Then, because agriculture production fell by half, there was famine accompanied by disease. At least another ten million, and almost certainly more, perished.

In his biography Malcom Muggeridge tells of a journey he made to the

Rostov area in March 1933 as a journalist for *The Guardian*. He described the famine that had devastated this most fertile area of Russia. "There is not only famine," he wrote, "but a state of war, a military occupation." Muggeridge goes on to say:

> **Stalin's collectivization of agriculture was...a general idea in a narrow empty mind, pursued to the uttermost limit, without reference to any other consideration, whether of individual or collective humanity. To be oppressed by an individual tyrant is terrible enough; by an outraged deity, as the *Old Testament* tells us, even more terrible; but Taine is right when he contends that the worst of all fates is to be oppressed by a general idea. This was the fate of Russian peasants, as it is, increasingly, the fate of all of us in the twentieth century.[3]**

It was indeed a great, swooping, voracious raid upon the human condition. It was all justified as necessary, as part of the process of creating an ideal. Neither did the fact that agricultural production fell by half cause substantial revision of concepts although it did give rise to massive, arbitrary *ad hoc* changes in day-to-day controls. But the concept of collective farming remained intact. Ideology took precedence over production.

The third epochal event was World War II. Here an external force impacted upon Russian agriculture. Much of Russia's most fertile lands were scenes of great battles. Productivity, which had risen slightly by 1941, was again severely hurt. Productivity fell again by about 50%. A great percentage, no one knows quite how much, of farm equipment, machinery, fixed infrastructure, transportation, and livestock was lost in the war. Farm manpower was gravely impaired when the war ended. In 1953 Stalin died and slowly thereafter Khruschev attained power. At first Khruschev endeavored to make the old model of collectivization work. By 1957 it was clear that, despite even introduction of virgin lands into the agriculture operation, the old collectivist model was failing. Khruschev recognized this and introduced yet another schemata.

Again it came right from the top. There never seemed to be any suggestion that the real agricultural change might come from the farmer. To the outsider it seems extraordinary that in the face of failure, little consideration was seemingly ever given to the man in the field as a possible progenitor of a better way. Nevertheless, if one looks at the matter from an ideological point of view, the situation becomes less strange. If one has given one's life to an ideological construct, one tends to blame failure on anything other than the construct. One can always find alternate scapegoats. That the construct could be wrong would destroy the supports of one's existence. Such supports one does not kick away lightly. A further reason why change from below was not instituted was that dramatic change had occurred in the human underpinning of Soviet agriculture. To a significant extent (and some say entirely) the old peasant structure had been destroyed. A form of peasantry did continue, centered around private plots, but the old Russian pattern of village and commune, territoriality and marketing, had virtually disappeared.

The most recently epochal event in Soviet agriculture came about in the late 1950's. Most commentators like to classify this as the gradual disappearance of the collective and the emergence of the State farm. The general idea was much the same, that of large holdings of land under state control. A certain relationship began to enter the picture. The procurement system was changed to that of money transactions. Workers were guaranteed minimum wages and pensions. Procurement prices paid for commodities (which hitherto were by fiat and entirely artificial) were increased, also by fiat. Machine Tractor Stations, centralized depots holding farm equipment, were abolished; this was a major progressive step. A significant effort was made, partially successful, to bring virgin lands under cultivation. In fact a new type of agriculture came into being, different in most operational aspects from the collective. Yields have improved and are coming close to world standards for comparable climatic regions.

Will the ideologues now be satisfied? Have the great swooping, voracious raids stopped? With regard for the human element involved, it is to be hoped that the answer is yes. In a system so centralized, however, enormous risks become inherent. After his fall, Lysenko was held responsible in some quarters for large scale crop failures. It was alleged that his method of plant conditioning was responsible. Whether the charge was well founded, is not the point of discussion here. Rather it is the vulnerability of a centralized system endeavoring to prove an ideological point. A change of direction is ordered from above and theoretically the entire agricultural edifice wheels in response. The order from on high had better be correct: otherwise disaster befalls the entire system and the nation itself.

Non-ideological systems are not so vulnerable to this kind of manipulation. Change in the U.S. comes differently. One or two farmers assume the role of guinea pigs, the entrepreneurs. If they succeed, others follow, often reluctantly. It is a less spectacular process, but it produces results without awful risks, to say nothing of loss of human life. Above all, each day in the U.S., about two million farmers are making agricultural decisions about relatively small areas of land with which they are intimate. Decentralization massively spreads the risk.

Ideology has affected Soviet agriculture in other ways. Presumably, if the Russian experiment is to be accepted at face value, the ideologue is less interested in the capital cost of the production mode than he is in the mode *per se.* Here is a great contrast with the West. If one does not care about how much money one expends, presumably one can do anything in agriculture. It is possible to produce orchids at the North Pole and polar bears on the equator if you are prepared to pay the cost. However, one would find oneself with very expensive orchids or polar bears.

The Soviet Union has made very high investments in agriculture, on a scale that no Western agricultural enterprises could survive without massive subsidies. During the 9th Five Year Plan the Soviet Union put 25% of its total national investment into agriculture. During the same period, the comparable U.S. investment rate was about 5%. The Soviet method of accounting for capital formation and the subsequent return on invested capital is different from that of the West. Nevertheless the fundamental principles involved remain the same. Capital has to be created. The Soviet people have to create their capital as does anyone else, by producing a surplus which is saved. Their process of doing this is through state control and direction, but the capital arises from the sweat of their brow, as does capital anywhere. Return on that capital is therefore just as meaningful in the Soviet Union as it is elsewhere. Someone, somewhere, has to pay a higher price for the product produced as a result of that capital investment.

Expressed as a percentage of private consumption expenditure on food, beverage, and tobacco, the cost to Soviets of their agricultural method is one of the highest in the world. A Soviet person spends up to 50% of his income on food, beverage, and tobacco. Cattle and hog prices measured at the official rate of exchange are nearly double that of U.S. and the yields of Soviet animals are less, another important cost consideration. These products, plus many others, are subsidized. In 1975, the Soviets subsidized meat and milk to the scale of an estimated $22 billion at official exchange rates.

In contrast the U.S. citizen expends some 17% of his income on food. Agricultural subsidies in 1973, including concessions on capital gains, were $900 million. From an agricultural viewpoint, these U.S. subsidies were less agriculturally than politically oriented. As usual with subsidies, there are inefficiencies, especially in the use of capital, affecting costs to the consumer. Under the system of overall state control the provider is not difficult to discern. It is the Soviet populace, whose burdens over the past 50 years have been considerable.

While the grand drama of Soviet agriculture was being played out, another intriguing, illustrative aspect of Soviet agriculture moved quietly along a clearly discerned path. No summary of Soviet agriculture, let alone one that attempts to relate ideology to husbandry, can pass without observing the remarkable phenomenon of the private plots.[4] These family plots of an acre or less are farmed by members of the collective. They form less than four percent of the total land farmed. At first glance they seem to offer the collective farm worker something to do in his spare time and perhaps a way to supplement and vary his diet. Under the system, the private plot farmer could also sell any surpluses he had in a public market. But this would be an incorrect observation. Numerous campaigns waged by Soviet authorities against private plots indicate that the plots have a significance that belies the tiny fraction they are of total cultivated area.

The plots arose as a key feature in the winter of 1929-30, during the drive to collectivize Soviet agriculture. The drive encountered peasant resistance, and nowhere more vehemently than by the peasant parting with his horse and especially his cow. It was the peasant woman who led the latter revolt and sporadic rioting that occured. Stalin referred to this contretemps with the collective farm women as "our minor misunderstanding about the cow." The misunderstanding led to events out of proportion to anything Stalin imagined.

By 1935 authorities had made concessions. They compromised with peasants by fully legalizing the plots. In particular, they guaranteed that a certain number of cattle might be kept on what was now a plethora of tiny family farms. Tension nevertheless remained, mostly brought about by the government constantly alleging that the collective farm worker pursued his private ends on his plots, chiefly those of making money, at the expense of work on the collective.

On May 27, 1939 Stalin launched a strong attack on the plots. He claimed that by general connivance they had been enlarged, the number of cattle increased, and peasant attention accentuated the importance of their little farms. New taxes and assessment quotas were laid against the plots. Committees were set up to check their size and return surpluses to the collectives. The campaign was labelled as anti-capitalist. Stalin's restrictions did reduce the numbers of cattle, something Soviets simply could not afford. What had happened was that by this time these tiny plots were producing such a significant part of total food output that they could not be reverted back into collective farm production.

In 1938 in USSR it was estimated that the plots formed 3.9% of the farmed area and produced 45% of total farm output. In one commentator's words, the plots had become "a giant dwarf." Collective workers received from the plots about half their cash earnings, most of their potatoes and other vegetables, and virtually all animal products. The plots maintained almost half of the nation's cattle, produced 70% of the meat, and 71% of the milk. Of critical importance to the non-farming sector of the Soviet economy, was the fact that these plots produced beyond their own consumption, about 20% of all marketed food supplies in USSR. Obviously this 20% accounted for a significant increment in peasant cash income. This extraordinary performance, it must be remembered, came as a *by-product* of collective farm work. The plots consumed an estimated 6% of the male labor output and 34% of the female labor output.

By the 1970's the plots had decreased slightly as a percentage of the land farmed, but they still produced 20% of the marketed food. One must note that the plots may have received benefit from the collectives not reflected in the data. For example, peasants received grain from the collective. This freed them to concentrate on more labor intensive operations such as dairying, and to labor on their own behalf. Other incidental benefits came from the collective such as use of some equipment, child care, and medical attention. The collective represented a kind of general bank from which the workers drew "loans" of this or that and from which they allegedly stole incessantly. This latter contention is strengthened by the constant charges of such stealing made in the Soviet press and by Soviet authorities.

The conclusion has to be that, no matter how they did it, these tiny plots produced a decisive portion of Soviet food supply, especially critical, as an ironic paradox, during the trauma of collectivization. Constant warfare by government against the plots also reveals their importance. They undoubtedly represented to the ideological mind a remnant of capitalization, a "survival of petty-bourgeois mentality." At the same time their contribution was such that they could not be abolished. They were so significant that here was one time ideology could not take precedence over production.

At this stage Soviet agriculture represented a hodge-podge of characters in ideological conflict and a system barely creaking along. Mosche Lewin, a distinguished commentator, summarized it well:

> **Thus the peasantry as plot owners were ideologically illegitimate, barely tolerated for the time being and deeply immoral.**
>
> **The whole so-called collectivized agricultural organization thus resembled a peculiar, awkwardly shaped cart with two huge wheels and one small, the one dragging the reluctant peasant into a pseudo-cooperative, the other (the Machine Tractor Stations) supplying them with machinery over which they had no say, and the third feeding them. At the same time the three disproportionate wheels, instead of really pulling in one direction, were badly coordinated and all too often blocked each other, each of them operating under different constraints and a different, often contradictory system of motivation. Without massive state coercion nothing of this kind could have existed for long in such form. It did of course continue to exist, but the results of such husbandry, not unexpectedly were extremely disappointing.[5]**

It was not surprising that the main problems soon were seen not as those of ideological dogma interfering with human creativity, but rather as problems of "re-educating the peasant." This is a classic ideological stance. The ideological dogma is never wrong. If there are difficulties, it is because of heterodoxy or ignorance. It is on these grounds that some unfortunate strata of humanity are bullied by an omnipotent state in an attempt to force submission. Great stress is also given to the future. Despite today's difficulties, it is said, the future remains glorious, for does not the ideological construct so proclaim?

Because of relatively short time span of human life in the face of great historical movements, the future belongs to later generations. Thus failure today calls for more intensive indoctrination of children so that they will become the appropriate standard bearers. It becomes important that these critical children be not contaminated by their elders. So the state considers it a duty to intervene in the raising of children. The presupposition is that by these means the tautology ultimately fulfills itself. Nurture of children is like that of agriculture, a complex matter based upon a web of human relationships. It responds slowly and cumbersomely to state instructions and controls. But the ideologue persists, indeed lapses into dogma. The ideology may even change in that it deteriorates, and this is not pretty.

Again to quote Lewin,

> **The conflict with the peasants was imposed upon them by political constraints and ideological predilection which were no more than an expression of an attempted rape of social realities.**

The parallel between Stalin's collectivization and serfdom, ostensibly abolished in Russia in 1861, has often been made. The peasant remained an inferior class with poor schools, few cultural opportunities, and little physical mobility, unless he was prepared to move to another collective farm or into industry. Later when they could, millions did just that. There was no social security system for peasants. There was poor housing, heavy taxes on the private plots, high prices for goods, and a system of state duties which townspeople escaped. Basically it was a system of forced labor common in concept and practice to that of the Russia of previous centuries. The collective system did not produce and for a very simple and basic reason, a reason related to agricultural productivity anywhere in the world. Productivity did not increase because the people involved, the farm workers, lacked incentives except on their little private plots. One can utter an obvious axiom. Where there are no personal incentives one has a stagnant agriculture.

Nevertheless, the West must realize that Russian agriculture became highly mechanized, with equipment based upon European and especially U.S. design. Even this significant and rapid mechanization of Soviet agriculture had little impact. Until the late '50's machines belonged to the Machine Tractor Stations and were parcelled out by bureaucratic procedure. Sometimes the procedure was corrupt and favoritism became the criterion. Agriculture cannot be efficiently operated in this way. The farmer must be able to relate the machine to the nature of the symbiosis. Plants have a biological time frame and weather can alter the best laid plan. The system obviously did not work; the peasants did not care for it, and this manifested itself concretely. As Lewin observed, "The Kolkoz fields were a cemetery for tractors."

The collective farm concept in conjunction with private plots did achieve one unsought return. Collectives gave the peasant something to rail against, if not hate. At the same time the private plot (which was also the peasant's home) gave him something to cherish. Here old ways and traditions were preserved. Here too was a physical security arising from animals, garden, and selling of surpluses in city markets. The system in fact helped hedge the cult of peasantry from the great attack being made upon it. The regime recognized this. Along with the poverty of collective farm production, this was reason for the changes of 1958.

In retrospect one marvels at what ideology hath wrought. From a successful agriculture under the NEP we witness the slide to a deterioration from the 1930's to the early 1950's. The slide was uneven, broken by a great famine at the beginning of collectivization. Problems ensuing throughout the period could have been devastating, had not gross production been boosted by development of new lands. Then we witnessed a virtual abandonment of collectivization and certainly of the theory that control of a particular farm was to be given to the workers at some future date. Instead we see advent of what might be called state farms, where the objective seems to be to bring operations as close to a factory style as possible. Yields have increased, and the system's productivity is not far behind the U.S. for comparable climatic regions. While all this occured, the private plots persist, playing their key role, feeding the Soviet people in a substantial way despite their tiny percentage of total farm area.

Yet the Soviet Union is still a grain deficient country. The Soviet diet is low in protein, especially animal protein. In 1974 in the U.S.S.R. all forms of protein comprised some 23% of the total diet, while grain and potatoes comprised 50%. In the U.S. for the same period, Americans consumed 38% protein and 24% grain and potatoes. In terms of meats the American consumed 251 pounds per person and in the Soviet, 108 pounds. Despite previous famine and overall terrors, most Russians were adequately fed. But they paid more for what they got than did their American counterpart, by a significant margin.

With their system it simply costs more to produce food. Nevertheless, there seems to be no significant change in Russian diet from the best of Czarist times to the best of Soviet times. Comparably, advances over this time span have been more than matched by almost any country in the West. Thus, besides asking what might happen next under the ideological constructs of the Soviet Union one can also ask, has it all been a necessity? Has attainment of reasonable agricultural productivity been worth the human pain and suffering?

One also comes back to another point, what is to happen next? Although data is sketchy, there are indications that morale of farm workers in the Soviet Union is low. One survey suggests that only 2% of young people on the Soviet farms wish to stay there.[6] If true, this is a staggering statistic. Certainly productivity of farm labor in the Soviet Union is low. Compared with the U.S., on collective farms growing grains other than corn (where U.S. manhour productivity is very high) in 1977 the Soviet worker took 1.4 man hours per metric centner compared to 0.3 for the U.S.; for cotton, 33 manhours in the Soviet Union to 3.6 in the U.S.; potatoes, 2.9 to 0.4; cattle, 50 to 3.7; hogs, 34 to 2.2 and for milk 9 manhours per centner to 1.3. These figures were for the period 1970-73.[7] As a result one farm worker in the U.S.S.R. feeds seven people, while in the U.S. one farm worker feeds fifty-four people. Thus we must ask again; was the journey really necessary, this long travail arising from an ideological construct? More importantly, does the travail have to be repeated by others who follow the ideological path?[8]

The question takes on a new dimension when we apply it to China. It is fair to say that since the coming to power of the Chinese Communist in 1949 people of the Chinese mainland have been subject to ideological pressure. Here, on the mainland, we find the world's largest group of farmers. Of China's 2,350 million acres, some 250 million acres as of 1978 are cultivated by an estimated 855 million farm people, although no one knows for certain how many farmers or farms there are. These farmers feed themselves and perhaps another 150 to 200 million persons. China thus farms some .25 of an acre to sustain each Chinese person. It can truly be said that under its present cultivation pattern, China is one country where population pressure upon currently cultivated land is real, though still substantially less than in Japan and many other countries.

Another way of expressing China's situation under its present cultivation pattern is that China must support about 25% of the world's population on approximately 7.8 percent of the world's currently cultivated area. Nevertheless, if it wishes to produce agricultural abundance, China's problem is about the same now as it was a century ago. It must obtain yields on each cultivated acre that are very high, at least as high as those in Japan and Taiwan. It must bring new lands into cultivation, as well. In this latter regard, a key factor especially in west China, is development of irrigation. Chinese leaders preceding the Communists, some emperors, Sun Yat-sen early this century, and later Chiang Kai-shek, understood the problem in much the same way. It was the Chinese Communist Party that found itself with the opportunity to do something about it.

Beginning in 1949, Chinese rulers divided their peasantry into four major categories. The first were landlords, who owned more than eight acres per household and did not work the land. Second were "rich" peasants, who owned five to eight acres and provided part of the required labor themselves. Then there were the middle peasants, who owned from 1.7 to 5 acres which they farmed themselves. Last there were poor peasants, who owned less than 1.7 acres and were largely tenant farmers.

In June of 1950 a land reform law was promulgated. Some 116 million acres were confiscated from landlords and rich peasants, and redistributed to some 350 million poor peasants. Scale of the operation was vast. Landlords and rich peasants vanished as a class. No one really knows how many were slaughtered. Estimates vary between 16,500,000 to 28,000,000.[9] A great deal of the slaughter was ritualistic as befits implementation of an ideological construct. Villages held public trials of the landlords and rich peasants, those unfortunate persons whose material aggrandizement was so little more than that of their accusers. Crimes, some real no doubt, others elaborately exaggerated, were publicly proclaimed. Passions ran high. The accused were invariably condemned to death. Sentence was usually carried out immediately. With communist apparatus in control, villagers were made to file past and perhaps beat the body with a club or even stick a knife into it. Here was a process identifying each person not only with the act, but with the regime. Later this process accompanied other communist-led insurrections in Southeast Asia, notably in Vietnam. There was also systematized slaughter in forced labor camps along with mass deaths through disease, fatigue and starvation.

These land confiscations were barely completed when the Peking government embarked upon yet another dramatic step. A simple system of agricultural cooperatives was formed; some ten million existed by 1954. Through these cooperatives the regime endeavored to make larger production gains through collectivization of agriculture. Lands, though privately owned, were joined together to make larger production units.

By the end of 1956 the situation changed again. Through centralization new cooperative units were reduced to 700,000. These new forms involved some 120 million families, or 95% of the total agricultural force. It was a massive consolidation, a real institutionalization of the collectivist principle. About this time there was yet another dramatic event. Except for tiny private plots, redistributed land was turned over to common ownership and workers received wages based only upon the amount of labor they provided.

In 1958 came yet another attempted transformation, the Great Leap Forward. This began, as far as agriculture was concerned, with the true commune movement. By late 1958 the 700,000 collective farms were consolidated further, this time into about 25,000 communes. Not only was the commune to be a production unit, but it also became the basic unit of rural government. Private ownership was eliminated. The commune operated as a double-edged weapon. It was not only to be a unit of government, but in doing so was to control all social aspects of peasant life. There were the famous communal kitchens, homes for children and the elderly, and a communalization of veritably all forms of the general activities of life. The more doctrinaire (and not only in China) hailed it all as a massive and rapid advance toward true communism. What was really happening was the advance of total control by the central government over China's masses, and concomitantly, a breakdown in farming practice.

Immediately succeeding the Great Leap Forward agricultural production, like industrial production, fell dramatically. At the beginning of the Leap, China's grain production had reached a total of 270 million metric tons. This was substantially more than double the very low base of 108 million tons in 1949 when China's agriculture, after nearly two decades of external aggression and civil war, was in a sorely distressed condition. By 1960, one year after the Leap, China's grain production fell from its high of 270 million tons to 150 million tons. This is an estimate because for 1960 the communist government imposed statistical secrecy. Ten years later it had not even reached its 1960 figure. A revision was obviously called for, and it came, once more.

Production brigades had already been formed, these being the basic production unit of the commune system. By early 1962 these had been broken down into smaller units, the production teams. The original 25,000 communes were increased to 74,000, but more importantly, production brigades totaled 700,000 units, while numbers of production teams rose to some five million. Production teams were made up of no more than twenty households.

Here then was a massive human ebb and flow. In space of less than a decade, Chinese agriculture had moved from a more or less mass system of peasant holdings to a tight, intense centralization. Rapid, large scale decentralization followed. All that was needed was for the twenty households to be given individual autonomy, and we should find ourselves back in China of the past. This was precisely what was happening, with variations.

Under the changed system, yields per unit area in China did increase, partly because of expanded irrigation. From 116 million irrigated acres in 1949 (rather a poor year to begin any statistical analysis) the irrigated area rose to 163 million acres by 1958, although doubt has been expressed that these figures are in fact true. There is no doubt that old, damaged irrigation systems were repaired, and new ones constructed. Current yields of rice in China at 1.04 tons of grain per acre are 38% of those in Japan. Agricultural yields in China are returning to what they were before China's gradual, but real collapse. Earliest indications that are pertinent suggest that in the early part of the 19th century, yields of rice were not too far short of those of today.

Under the ideologies, the entire system nearly collapsed. More than that, the impending collapse was so obvious that the regime had to take vast, swift action. The speed with which the regime disestablished communes testifies to the seriousness of the situation, as does the continuing attitude of the government toward agriculture. By 1969 Chinese peasants, faced with the disaster of the communes, were openly stating that the regime had lost the mandate of Heaven to rule. One sure sign of such a loss of mandate was traditionally a disaster in agriculture along with other signs arising from natural phenomena. As with farmers anywhere, the enormous loss of productivity was observed by them on a daily basis and could not be hidden by any amount of rhetoric. Agriculture came to take prominence in public discourse.

By time of the Ninth Party Congress in 1969, agricultural priority became so important that the new national policy became to support agricultural needs by industry, instead of agriculture supporting industry. Previous dissipations of peasant effort by such extraordinary demands as producing iron from backyard furnaces were halted. Industrial production emphasis was switched from the Great Leap Forward phase of production of "non-essential items," and re-programmed to manufacture products used directly in farming. Mao Tse-tung himself offered an "eight character" design for scientific farming which was entirely based upon small units and small unit personnel operations.

At about this time, youth from universities and middle schools, as well as bureaucrats, were sent into the country to work with peasants, to see reality of things. This idea may have a more general application and significance for countries outside of China. Skilled workers, usually artisans, were also dispatched to farms to help achieve self-sufficiency. Thus modern Chinese agriculture, set in motion by gigantic massacres and revolutionized by the commune system, underwent another transformation as it returned to a near traditional mold.

Return to a near traditional organization is a fascinating issue especially as this happened during a time when China was, if one accepts the ideological rhetoric, to be kept in a state of permanent revolution, wherein China was to produce a new Chinese man, cleansed of all the evils of the old ways. For the new production units *were* traditional. Of that there can be little doubt.

To these traditional forms have been added new techniques. There has been some improvement in irrigation. New seed strains have been introduced. Indeed, indigenously developed strains became available from Chinese research at about the same time similar strains were developed in the West. Chemical fertilizer and pesticides have been introduced on expanding scales. Though the level of mechanization of agriculture is still extremely low, this too has improved somewhat. What happened after failure of the commune system was that China achieved a low level technological mix, between a traditional agriculture and the minimum of new agricultural sophistications.[10]

Emphasis on the production unit represented a return to a form of agricultural mobilization with which Chinese peasants were familiar. Chinese society has always been highly structured; by "always" in China, one means at least two, and probably three, millenia. From the beginning of the Empire in 221 B.C. until the China of the present, the idea of collective action, especially by small groups, has been a cultural predominant. This applies especially to agriculture, but also carries over into virtually all aspects of Chinese life.[11]

However, under the communist system of the Great Leap Forward, collectivization went too far, even for the Chinese. In fact it contravened Chinese tradition. To take food produced away from the tiller and then give it back to him in a communal kitchen represented a breakdown in society similar to the historical breakdowns that arrived with war — political turmoil, floods and famine. The peasants also knew (certainly before the ideologues) that production was falling under the Great Leap. Here, in particular, Chinese peasant sensitivity was acute.

Nevertheless, the Leap was swiftly abandoned and the smaller production teams formed. The most significant aspect of this new construst was that production units retained sufficient food for themselves; they got first draw, so to speak. As an added incentive, they retained the bulk of the increments in income arising from any improvement in productivity achieved. The private plot system was accentuated as an added incentive, an accentuation that continues today. The posture of relatively small teams cooperating toward common agricultural purposes is so traditional as to be indistinguishable from long established and ancient patterns. Only the modern water pump, fertilizers, pesticides and the like are different.

Otherwise, there are two main features: (1) the tiny collective, or production team, at work, similar to the traditional Chinese Ching T'ien or "well fields" system of remote antiquity; and, (2) the system of "equal fields," or Chun T'ien, also ante-dating the Mao Tse-tung era by two millenia or more.[12] The former was the norm for very long periods in Chinese history. When political breakdown was sufficiently widespread for people to question the current dynasty's right to rule by the "mandate of heaven," then one of two things would happen. Either revolution would bring down the dynasty or dynastic rulers would move effectively to right things by invoking renewed social equality under a system of "equal fields."

In 1949, after the enormous breakdown occurring over the previous century, equal distribution of land was Mao Tse-tung's rallying cry. Some twenty years earlier the Kuomintang under Chiang Kai-shek had tried, under conditions of civil war and external aggression, to do the same thing. Peasants responded to Mao's call as they had traditionally to any new ruler who obeyed the rules directed at restoring harmony. Expressed in simple terms, Mao took land back from the peasants under the commune system. The commune system failed, for it was not a harmonious relationship. For example, its products went first to the state. Sufficient foodstuffs were then to be given back to peasants in the com-munal kitchen. Women were given equality, the equality to work in th fields alongside men (which, in fact, peasant women had always done Their children, traditionally the wife's first duty, were to be cared fo within various commune institutions. Here women balked. Old people, grandparents, had traditionally cared for children when parents were in the fields. This custom is still followed over great areas of Asia. It is at once a form of social security and an assigning of importance to an elderly relative.

But a more important issue pertained and needed a positive response. The peasant's fundamental and never-changing question was, "What is there in this for me?" When he received no answer, he "leaned back on the hoe." Here was the primary cause for poor production. The plots distributed in 1950 and shortly after were too small to give reasonable living standards to the farmer after exactions of the new landlord, the state, had been met. Later, with communes, the state deliberately sought to destroy traditional family and village life. Dormitories, segregated by sex, were a particular example and probably represented a crude state-sponsored birth control device. Also, the peasant could visibly witness the productivity of tiny, private plots and could make the obvious comparison.

Ideology remained, but in a very special form. To Chinese communists, capitalism is not so much a method of material production as it is a state of mind. Capitalism represents persons who would aggrandize themselves at the expense of others, who demonstrate greed, lust, and disdain for their fellows. Thus, reason Chinese Communists, eradication of capitalism is an inner thing. It is personalized in the sense of self cleansing, of self-regeneration. There is no counterpart of this attitude in the Soviet Union. Ideology in China, therefore, did not take on only a physical, didactic construct; instead it became a personal, flexible doctrine.

Perhaps, on the other hand, Chinese communists did not have much choice. By every measurement, the Great Leap Forward was a disaster. If for ideological reasons communes had been retained, there seems little doubt that the resulting chaos would have reached that degree where the ancient notion of a lost Mandate of Heaven might have resulted in an overthrow of the regime. There is some evidence that mutterings about a lost Mandate were already abroad before the reversal came. Ideology as a productive construct therefore failed. But then, ideologues are often impractical fellows. They assume that practical implementation will follow smoothly as a logical extension of the idea.

The question that arises for more practical men, as they proceed to implement a concept is, "will it work?" This often does not bother the ideologue. The test comes when practice becomes failure, but the true ideologue is not bothered by this. Failure occurs not because of the idea, but for some other reason. Thus effort must be intensified to make it work, or the alleged impediment must be exorcised.

Some of this happened during the Great Leap Forward. The Chinese reacted to failure pragmatically. With enormous vitality and considerable flexibility, they turned to their past in the productive sense, and relegated ideology to something near spiritual. Even that has probably gone. The name given to the notion was Maoism, and Maoism died with Mao. This has allowed the Chinese to address themselves to practical problems once again more directly.

It would be wrong to assume from the foregoing that by this retreat Chinese agriculture on the mainland has solved its problems or that it is an effective agriculture. It has a temperate climate and has yet-untapped agricultural resources, especially in the west. Still, Chinese communist agriculture limps along with a reasonable modicum of productivity, a small expansion rate that just about keeps up with population growth. It bears a bedevilling burden of petty bureaucracy, shocks of cultural revolutions, and the ever-present threat of yet another change born of some ideological notion in Peking. As one contemporary observer of the Chinese scene noted, China shows "a maximum of unfreedom with a minimum of efficiency."[13] It is hardly a model to be followed, and once again, one can marvel at ideology's impact on world thinking, especially among the young who often cite China as a success.

We must now turn to the ultimate paradox. Despite rhetoric, when it comes to agriculture, the mainland Chinese are far less ideological than it first appears. They made their great, swooping, voracious raid in 1959. Being pragmatic people, when failure was obvious, they immediately turned back to the traditional capacity of Chinese people to cooperate in small groups. This tradition runs deep in the Chinese family, with its intense vertical loyalty and its equally intense, though different, horizontal loyalty.

One hundred and ten miles across the Formosa strait is another China, the Republic of China on Taiwan. The Chinese on Taiwan have developed a highly successful agriculture, perhaps the finest in the world. It is also a structured operation based on combined centralization and decentralization. Guidance and promotion from government goes along with a wide-ranging series of Farmers' Associations, which function as cooperatives. We have already noted the nature and significance of these. Beyond the cultural stimulus Chinese have *had* to cooperate, and more so today. Where irrigation was the basis of their agriculture, cooperation became essential to provide fair shares of water for all. Today, individual small plots cannot support expensive machines, so smaller plots are consolidated into larger units. The same might be said of purchasing and marketing in what is becoming, in Taiwan, an increasingly complex commercial environment. Here, one again sees the old Chinese capacity for cooperative action at work. In terms of straight operations, a Chinese team in Taiwan would likely feel quite at home on the mainland. But there are significant differences which reveal the debilitating influences of ideology.

The differences arise out of different ideological constructs and are great portents for the future of the societies concerned. As in feudal times, land on the mainland belongs to the state. In Taiwan, land had been passed over to individual ownership; some 94% of all arable land is owned by peasants. An argument is not being made here for the virtues of individual ownership of land. It, too, has problems, notably those pertaining to inheritance. Taiwan has yet to face this issue, if only because younger members of farm families are easily absorbed into growing industry.

Rather, the argument centers around the short and longer term productive efficiency of state as opposed to private ownership and the impact of either on people. If the state owns agricultural land and at the same time attempts to plan and control agricultural operations, then for the system to work, the state constantly has to play the role of director. No invidious comparison is intended when it is observed that prison farms in the U.S. offer a parallel. Some prison farms attain reasonable agricultural productivity. The problem is that for this system to go on working, the central prison authority must always be in totalitarian control. Further, entrepreneurial talent must also come from the central authority. If that authority loses control, prisoners either will not work or flee if they can. Already Chinese authority on the mainland is coming face to face with this problem, just as has the Soviet Union. But a central authority in a large state or area cannot exercise control as can be done in a small prison.

Mainland Chinese have from time to time declared their shortage of entrepreneurial skills, and one suspects that it will become an increasing problem until incentives can be given beyond that of being ideologically pure. Chinese on the mainland have also been aware of the problem surrounding centralized control, as well as the lack of entrepreneurship. It was no accident that they limited their production teams to about twenty families and related these families to proven traditional marketing and social units. People in the team knew each other and were often relatives. They could coalesce as a unit around traditional structures for cooperation and for exerting the usual pressures to get work done.

Another significant change on the mainland, which has been further publicised since the death of Mao Tse-tung, is the support and encouragement given to private plots, these being introduced again after 1959. Production teams have now been asked to give over some 7% of farmed land for allocation to individual families, wherein they may have a private garden. These plots are primarily devoted to production of fruit, vegetables, tobacco and herbs, hogs and chickens. Families are encouraged to sell surpluses in local markets. All indications are that despite extremely strict regulations, private plots have been production successes. The question now arises: how much further can these kinds of incentives go in communist China without coming against ideological constraints? How far will the mainland go in following Taiwan? How much further can they go without leaving themselves open to charges of "revisionism" or "capitalist roading?" Here probably lies the answer to China's future agricultural success. How to take yet another step away from ideological posturing and toward agricultural productivity.

The Taiwan experience may indicate that the next step would be: retention of old forms of cooperation, plus the incentive of individual decision-making, even individual entrepreneurship. It does not have to be individual ownership of plots because even that is alien to China to some degree. The traditional idea of the Emperor as land holder and the peasant operating land in perpetuity is not totally abhorrent. The state can play the same role if it must. This might be called a bending of ideology. Thus the question of the future returns again to the issue of ideology itself. If ideology returns to the fore, if the communist system is accentuated, progress will probably slow or even halt again. It is now for Chinese communists to decide whether their "new man" revolutions will continue seeking out some ill-defined "new man" model to which people must be pushed regardless of their past or present aspirations.

But the demand for a fast changing world dies hard in ideology. The ideologue's mind is bedevilled with the notion that he is the midwife of the new, offering to mankind optimization of his condition. Thus, the true ideologue not only seeks continuously, but he seeks in faith — a faith just as real, passionate and devout as that of any early Saint. Yet the whole process may be much more tawdry than this, little more than a continuing process in human self-deception. It might be better to have left it alone, rather than to pursue this seemingly never-ending goal of remodelling people. In his usual iconoclastic style C.S. Lewis summarizes the issue:

> The demand for a developing world — a demand obviously in harmony with the revolutionary and romantic temper — grows up first; when it is full grown the scientists go to work and discover the evidence on which our belief in that sort of universe would now be held to rest. There is no question here of the old Model's being shattered by the inrush of new phenomena. The truth would seem to be the reverse; that when changes in the human mind produce a sufficient disrelish of the old Model and a sufficient hankering after some new one, phenomena to support that new one will obediently turn up. I do not mean at all that these new phenomena are illusory. Nature has all sorts of phenomena in stock and can suit many different tastes.[14]

In the meantime, the Chinese farmer must go on. Decisions need to be made each day. Planting, growing and harvesting cycles remain. Needs of animals must be met — and of soil. Weather may grace or refute. In his mind, the Chinese farmer looks at these and then his relationship to these factors. Like any one in a similar situation he seeks fairness, evenness, and a reward for what he does. This is *his* ideology.

In the world beyond China and the Soviet Union, we see either no ideology at all regarding agriculture (and that seems to be invariably so with successful agriculture), or we see a very weakened kind of ideology at play. The approximately two million farmers who make decisions in U.S. agriculture could fairly be described as a-ideological and we have noted their operations go remarkably well despite what some moderns would term this deficiency in their attitudes. By and large, the same might be said of farmers of Western Europe, Japan, Taiwan. I can testify that farmers of Australia and New Zealand make a wide intellectual detour around ideology and its rhetoric. At this stage in the argument another question is often introduced. What of those impoverished peasants of the Third World — do not they need a motivating ideology to lift them out of their agricultural torpor?

In 1948 Burma regained its sovereign independence from Britain. Burma had suffered greatly in the war with Japan, 1941-45. Before that time it had been the largest rice exporter in the world, with annual shipments reaching up to 3 million tons. The Burmese had cultivated wet rice long before the British colonization, which began in 1886 when Burma, after losing the third Anglo-Burmese war, was annexed as a province of India. Under stimulus of British-created markets in India and later in Southeast Asia, notably Malaya, there began a massive population shift in Burma. It comprised peasant farmers moving mainly into the flood plains of the lower Irrawaddy valley. Most historians regarded British economic intrusion into Burma as socially and economically disastrous. This British stimulus was in truth an accidental by-product of their overall Asian ventures. Nevertheless, the vast Irrawaddy Delta was basically cleared of its mangroves and became rice fields. This began by the 1850's. The price of rice for export had doubled by 1890 and this was the root cause of migrations into the delta and its development.

British efforts in this development were haphazard indeed. Wholesale trade in rice, including exports, was tightly controlled by a few British firms. Chinese and Indian traders controlled retail trade. The newly instituted banks, also British, were not involved in agricultural development; they did not make crop loans. These essential services were provided instead by Indian moneylenders, who had arrived on the coat tails of the colonizers. Nevertheless, export economy in rice grew in 75 years until, prior to World War II, Burma was the world's leading rice exporter.

During and immediately after the war, the rice export trade collapsed. Independence came in 1948 and in a decade the Burmese economy had largely recovered. Rice exports again hovered between two and three million tons per year.

After a series of political disturbances, General Ne Win emerged in 1962 as the leader of a revolutionary council which proclaimed a "Burmese Path to Socialism." Land had already been formally nationalized shortly after independence, but this had not been implemented to a significant degree. After 1962, the rest of commerce and industry were also nationalized. As a seemingly minor adjunct of nationalization, another event occurred which had immense significance. The activity of Indian moneylenders was abolished, in the popular belief that socialised agriculture did not need this private intrusion. Moneylenders were indeed a rapacious bunch. If anyone can ever generalize about hatred, the bete noir of the Burmese peasant was undoubtedly the moneylender. But even socialised agriculture needs credit and capital. This the moneylender had provided. They went. Nothing took their place.

The peasants reacted as was to be expected. They grew sufficient rice to satisfy their own needs. They no longer invested in credit to grow surplus for export, for such credit no longer existed. Not only did Burma recede from being the world's largest rice exporter to a status of no exports at all, but its cities were faced with escalating rice prices and even with rice shortages. During the 1970's, however, there was a small surplus for export, some 164,000 tons. During this time the world market for rice was depressed, and many Burmese offer this as an explanation. But neighboring Thailand, hitherto second to Burma continued during the same period to export in excess of two million tons per annum, despite these difficulties.

During this massive interregnum in Burma's post-war export drive, there was much rhetoric at home and abroad regarding the uniqueness of Burma's "Path to Socialism." But rhetoric, no matter how pleasing, does not grow much rice. Perhaps Burmese socialism was more rhetorical than ideological. Ideology, however, was the root of it all, exacted an inevitable debilitation of Burma's agriculture.

This may also be the case elsewhere in the Third World, where one finds ideological rhetoric. Sri Lanka turned down a World Bank loan to build a dam because the repayment terms represented "neo-colonization." Perhaps the repayment terms were indeed unfair. It would have been enough to have argued the case on those grounds, rather than in pseudo-ideological rhetoric. Tanzania is approaching its vast agricultural potential with all organizational forms of both Soviet collectives and Chinese communes. Mozambique appears to be heading the same way. These forms, however, may turn out to be little more than cosmetics, and a far cry from an ideology in action. The issue now is to summarize; to ask what it has all meant, this lusting after ideology. We must ask again if it were all necessary. Must ideology again make other voracious raids upon humble farmers?

First, we should beware of making ideology a battleground between communism and non-communism. One hesitates to say communism and capitalism for it would be hard to translate capitalism into a single coherent system, let alone a self-contained ideology. Communism is important to this study because it impacted as an ideology upon agriculture in vast measures in two great countries.

No doubt the future holds other ideologies. Communism happens to be the one currently in vogue and on a grand scale. Any other ideological candidate is at this time latent. But the situation does prompt the next question, one putting ideology into a constructive context. Can man live without a set of ideas to guide and inspire him? Outside of a motivating ideology, is not man's motivating force merely that of lust, greed, and envy? The question has been asked many times. It has also been noted that ideology in its modern forms, Fascism, as well as Communism, took hold when the spirituality and authority of the Church entered a period of decline. Perhaps what might have found expression in religious observance came to express itself in various forms of political belief. In other words, is it a truth that man craves a belief constructed in absolute, even dogmatic, terms to regulate and shape his mind and actions? Upon the answer to this question may hinge the future of world agriculture.

There is an alternate mode of thinking which well suits the farming mind. It is simply to take a pragmatic view of things while leaving oneself open to an improving future based upon one's own actions. Here we do not deal with closed abstractions, but rather seek the prudent. We can be well guided in this, for we have not only our own experience, but what we may obtain from others. We have tradition. Tradition suggests that time proves a methodology or a viewpoint simply by trial and error, by the adjustments of our forebears.

I have watched the observance of many strange customs in many parts of the world. Most were based on good common sense and an amalgam passed on by many people, and the observance of custom gave psychic health to the participant. Stress was absent. Traditional institutions may form an antidote to ideology. These may range from the Constitution of a state to the traditional construct of a family. In these operating arms, non-ideological communities tend to obtain flexible yet acceptable roots.

But more is needed. What has been stated is the construct of a traditional society where, in many situations, not only people's minds are constricted, but their agriculture also. To the pragmatic structure we must add another dimension of critical proportions, if not of ultimate significance. We must add the exploratory and imaginative mind. In agriculture we can base this on two broad themes; empirical observation with its subsequent judgements, and theoretical science. Here is a future-oriented, new concept, new method dynamic that traditional societies lack. Here we have a construct, but not the closed construct of a tautological ideology. In the dynamic of empiricism and science, we seek one primary objective, that of conquering ignorance, and it is suggested ideology is a special form of ignorance. This is not a new concept but an old one.

> **Sin is not the violation of a law or a convention...but *ignorance*...which seeks its own private gain at the expense of others.[15]**

From the conquest of ignorance, all ignorance, we advance to lesser ignorance. The ideologue basically presupposes that the ideology tells him everything, if he can but understand. It tells him that all his work, practical and theoretical, is proof of what ideology already knows. It is for this reason we find little originality of thought from within the ranks of ideologically oriented intellectuals. For this reason, research and thinking become a means to a political end. It is why Lysenkos arise. Paradoxically, Marx himself, who frowned on ideology, put it rather well except that he spoke of capitalists instead of ideologues. "The same men who establish social relations comfortably with their material productivity produce also the principles, the ideas, the categories, comfortable with their social relations.[16] Today we have to change the genre slightly. It is not capitalists by their material productivity who categorize, but intellectuals through their mental productivity. To accept these notions of continuing conformability to a doctrine that is not seriously questioned, is in fact to be a traitor to the human cause...to cease looking for truth.

Man needs to search, unknowing everything of his future, yet knowing much about where he has been. More often than not, he refuses to learn from his past, but the past is there to help analyze his present condition. An open, analytical mind is needed for this task, whereas the ideologue is proposing a stasis, a slowing or stopping of the articulation of ideas. One of the great paradoxes of Marxism noted by many and arising directly from Hegel is: why should the dynamic dialectical process, the explanation of history itself, suddenly come to a halt with the advent of Communism? In the ideologue's mind, however, stasis is reached when he embraces and then comes to master the doctrine.

In practical terms, ideology in power, this mind set in control, exacts a very high price from productive and thereby creative sectors of a nation. We have noted that ideology can be expensive in terms of human life. To bend a nation to one's will by killing dissenters is unfortunately a feature of the modern age. Ideologically it is justified. Ideologically, ends justify means.

It is also expensive of resources. We have noted that the Soviet Union spends five times more on agriculture than does the U.S. Further, despite this investment the Soviet Union produces for its people an inferior diet, especially if measured in protein, and often cannot meet its gross demands for grain. But what does this matter if one's goal is to bend agriculture into an ideological shape rather than a productive one? To the ideologue, the point of resource abuse is academic if not trifling. Yet we live in a world of quite finite resources.

Ideology in action is expensive of human time. It demands immediate change in human attitudes. Under these pressures human effort becomes that of psychic trauma and physical disruption. One can only ponder the manhours spent in China on their ideological constructs; trying and executing landlords, the Great Leap Forward, and the Cultural Revolution. One ponders the time and energy expended by millions of people on the "Little Red Book" and other forms of Maoist indoctrination. For farmers this becomes especially pertinent, since most indoctrination took place after sunset, after a day in the fields beginning at sunrise. And when that hullabaloo was over, one still had to come back to the practicality of plowing, planting and harvesting, much as one had always done. In the end, one must produce results; not merely psychic exultation. Time, especially in agriculture, has quite fixed dimensions. One cannot march and shout when it is time to till, time to plant, time to reap — unless one is prepared to pay the price.

Ideology is also expensive of human intelligence. It is incorrect to collectivise intellect. When one tries to, a likely result is one intellect ruling the collective. Intellect, when at its best, is personal and individual. Creative acts do not as a rule emanate from groups. The task is to free, not bind, intelligence, There is good reason why the Chinese complain about lack of entrepreneurship. The condition will persist until individuals are given an environment in which they they can act freely. An environment of freely-acting individuals does not fit well with ideologues. To the ideologue, thought of the American situation largely repels, where two million farmers have the right to make their own decisions, on their own. To the ideologue, this resembles not only formlessness and ideological confusion, but also a threat to ideological control. The ideologue cannot agree with Nietzsche that "out of chaos comes a shining star." Yet this is exactly what has happened with two million American farmers making their own decisions.

In the end, ideology is expensive to human progress. Ideology has a recklessness with history. It is happy to throw away what we know and often demands just this. It is like entering a cage of hungry lions. Beyond the likelihood of being devoured, it tries the patience and saps the vigor of millions of ordinary people who would rather be doing something else. But the vigor of an ideology in action must be acknowledged. The vigor of ideology in the Soviet Union during the Civil War; the overthrow of NEP; collectivisation with its destruction of people, its famine and recovery; during the change to state farms — it is all a drama of overwhelming dynamism.[17] The same can be said of China, especially regarding the Great Leap and rapid recovery from this policy. Was it all necessary? Was it human progress? Or was it waste, a totally misplaced effort, a misuse of an enormous explosion of human vigor? If we measure human progress in terms of improving agricultural productivity, we know from all manner of alternate examples that we simply do not need an ideological framework to obtain productivity. Indeed, ideology is not needed in agriculture, for farmers as persons are better off without it. Their a-ideology is healthy for agriculture.

And now in 1979 China engages us again. The disturbances caused by the "gang of four" following the death of the "great helmsman" was short lived. China now seems quiescent even though the vast majority of cadres and party members are unreconstructed Maoists still embracing the idea of romantic revolution. This is in contrast with the seemingly pragmatic mood of the new regime. They ushered in yet another successor to all those stages of convulsion each coming in swift succession after the slaughter of the early 1950's and the Great Leap Forward of 1958. This new stage has been designated the "four modernizations"; modernization of science and technology, of defense, industry, and agriculture. The new leaders of China, in direct contrast to all of their predecessors, tell us that China is a poor and backward country. The Chinese peasant in particular is singled out for criticism regarding his low productivity. Why he should produce so poorly should however be no surprise to anyone except the biased, the credulous and ideologues. The new pragmatism is best summed up by its most voluble spokesman, Vice-Premier Teng Hsiao-Ping, "The color of the cat does not matter: The question is can it catch mice?"

The advent of the "four modernizations" is a massive indictment of thirty years of Maoist ideology in action especially in agriculture, that area where China had the longest tradition and experience upon which to build a modern structure. The changes contemplated, such as tens of thousands of Chinese students studying in the West; heavy purchases of Western industrial technology; the resurrection of yet another area of Western dominance into a premier role, that of science and technology; the Westernization of Chinese education curricula and the search, oft repeated by Teng, for "a kind of socialism Chinese peasants will like," stands Maoism on its head. Or so it would seem.

The validity of the rhetoric has yet to be tested operationally. The key test will be in agriculture. If China cannot activate its six to seven hundred million peasants to produce the kind of sustained surpluses, which China is capable of producing, and needs to support a sustained industrialization, any other sphere of modernization will ultimately be dragged down by the sheer massiveness of the political, economic and social cost of rural poverty. Thus China stands at a crossroad and may be facing yet another convulsion, another conflict between ideology and production.

As portrayed in this book, successful farming and Marxist ideology are in direct conflict. Successful farming comes when the ethos, the farming ethos as portrayed in our first chapter, is optimized. It is this ethos in its practicality that is the antithesis of Marxist theory. And the antithesis is practical. The new regime on the Chinese mainland can see the ethos in action, amongst their fellow Chinese, 110 miles across the Taiwan straits in the Republic of China. This ethos will work equally well with Chinese on the mainland. The question becomes will the mainland Chinese, pragmatic as they may sound, embrace it? How can they Taiwanise the mainland: The reverse is to happen. Yet this is the nub of the new pragmatists' dilemma. They must so a-ideolize their agriculture, not merely to optimize production but also have their peasants per medium of this new productivity assist with the other modernizations as was the case on Taiwan. Could they alter their rhetoric to make our ethos the "kind of socialism a Chinese peasant likes?" A true pragmatic might!

A fascinating scenario is unfolding on mainland China in terms of what has been written in this book. It is as dramatic as the difference between the beginnings of a quiet but steady economic growth for the world's most populous country or conversely the certainty of continuing economic stagnation and an increasing human and technological redundancy. The latter situation will almost certainly bring about another convulsion. This group is likely to be ideological purists seeking to resurrect the true faith after failure of a heresy. Thus a new round of ideologically borne suffering may again lash the Chinese people and farmers most of all.

That of being a-ideological in farming, is of course somewhat tame by comparison — dull in fact. The relief from dullness is a certain chaos of individualism and an inability to anticipate what might come next in the sense of an improving and changing status. Neither can one in the a-ideological state obtain that exaltation of reaching for utopia. Instead, especially for the agriculturalist, one falls back on the prosaic task of tending one's garden. The only satisfaction one gets from this a-ideological state is that of contributing a small amount to real progress.

In the meantime, those farmers living in states having an ideological construct face a quite different problem. They must hope that they, their families, and the lands they toil on, will not once more be subject to yet another great, swooping, voracious raid.

FOOTNOTES

1 Emil L. Fackenheim, *Encounters Between Judaism and Modern Philosophy* (New York, Basic Book, 1973) p. 21.

2 E.H. Carr, *Michael Bakunin* (New York, Vintage Books, 1937) p. 48.

3 Malcolm Muggeridge, *The Green Stick: Chronicles of Wasted Time* (Glasgow, Fontana/ Collins, 1975) Vol. 1, pp. 286-287.

4 For succinct information on the private plots in Soviet Russian agriculture I am indebted to my colleague at the Woodrow Wilson International Center for Scholars, Dr. Moshe Lewin, University of Birmingham, England. The following pages represent a summation of a longer paper prepared by Dr. Lewin and which will be part of a multi-volume work on *The Social History of the Soviet People.*

5 *Ibid.,* unpublished paper, p. 7.

6 Supplied by Robert Lewis, Fellow, WWICS, NARODONASE LENIYE (Moscow, "Statistika," 1973) p. 38.

7 Douglas Diamond, *U.S. and USSR: Selected Indications of Agricultural Activity and Production.* Conference on Soviet Agriculture, Kennan Institute, Washington, D.C., November 16, 1976.

8 In addition to general reading listed in the bibliography, especial help was obtained from the following papers: D. Gale Johnson, *Theory and Practice of Soviet Collective Agriculture,* Office of Agriculture Economic Research, University of Chicago, Paper 75.28, Dec. 1975; D. Gale Johnson, *The 10th Five Year Plan, Agriculture and Prospects for Soviet American Trade; James R. Miller, Models of Soviet Agriculture: The Soviet Case in Historical Perspective.* Notes for a Talk on Future of Soviet Agriculture, Kennan Institute, WWICS, Nov. 1976. Marina Menshikova, *The American Way in Agriculture and its International Significance,* Paper for 1976 International Conference, "The United States in the World" Sept./Oct. Smithsonian Institute, Washington, D.C.

9 U.S. Congress, Senate, Committee on the Judiciary: "The Human Cost of Communism in China," 92nd Congress, 165 Session, 1971, p. 16.

10 Probably the most up to date survey of contemporary Chinese agriculture is Benedict Davies, *Making Green Revolution: the Politics of Agricultural Development in China* (Rural Development Committee, Cornell University, Ithaca, 1974.)

11 For example, an overview of the Pao-chia system is given in Edwin O. Reischauer and John K. Fairbank, *East Asia: The Great Tradition,* (Boston, Houghton Mifflin Co., 1958) pp. 28, 159, 206, 374.

12 I am deeply indebted to Father Joseph Sebes S.J. of Georgetown University and a Fellow at the WWICS whose scholarship on China is better than mine and who stimulated and broadened my own conception of the continuum of the Chinese tradition under communism.

13 Edward N. Luttwalk, "Seeing China Plain," in *Commentary,* December 1976, pp. 27-33.

14 Quoted by Jeffrey Hart, "Ideas in Culture," in *Imprimis* (Hillsdale College, Hillsdale, Michigan, Vol. 5, No. 11, Nov. 1974), p. 3.

15 *The Bhagavad Gita,* translated by S. Radhakrishnan (New York, Harper Bros., 1948), p. 244.

16 Karl Marx, *Poverty of Philosophy* (London, M. Lawrence).

17 Economic Research Service, U.S. Dept. of Agriculture, Foreign Agricultural Economic Report No. 92, *Agriculture in the U.S. and the USSR,* pp. 1-18, excellent short analysis of U.S./USSR agriculture, January, 1977.

AN INHIBITOR: POLITICS 8

In assessing the environment for inducing and recognizing creativity, we have noted the inhibiting effects of certain kinds of internal instabilities, and international tensions, and of political ideology carried to the lengths of a dogmatic closed cycle construct. These aspects of the way contemporary man goes about his affairs, of the environment within which men must work as farmers, are manifestly about us. They are, however, but two aspects of a much wider whole, this overall human environment within which we operate — farmer or otherwise. Regarding political ideology, we cursorily examined how people think, especially people in power. This is important, but so is the corollary. It is not only what people in power think, but what they do that creates an environment. There is often a gap between what people think and what they do. Thus it becomes necessary to go beyond ideology, to venture into the action aspect of everyday political life, which may or may not be political ideology in action.

We must, at this stage, provide a very general definition of politics. Politics is conceived to be more the control and management of people living together in a society or state, rather than the form, the organization and the management of the state. Form, organization and management can be reduced to written documents, constitutions and legalities, and even to charts. These often belie what actually happens to people in a political system. It is this that interests us. Again in this overview of planetary agriculture, we must look at this situation in general terms. We will, however, notice commonalities, and many of them. Again, concentration will be on the Third World, the area of greatest need and also an area where politics and its impact on people is as real as in seemingly more sophisticated regions.

The role of politics has changed. The public sector has now become a dominant feature of virtually every nation in the world. In some, in economic terms, there is little or no private sector at all. Even in "free" societies it is government, with politicians at the helm, who have arrogated unto themselves the basic control and management of societies. This also applies to agricultural development, for which politicians have now assumed responsibility. The needed sponsorship will be government sponsorship... political sponsorship, if you like.

Not only that, in most nations no alternate sponsorship will be permitted by political leaders. Wittingly or unwittingly therefore, political leaders throughout the world have put themselves out front. Increasingly, they shall be expected to deliver. Wittingly or unwittingly, they have chosen a lonely position. Agriculture is an area where performance can be fairly accurately measured. As we shall see however, the role politicians have assigned to themselves, created by their promises or ideologies, is somewhat spurious. In the end agricultural development must come from individual farmers. In terms of practical results, politics and polititians can at best play a small but important sponsorship role in creating the environment needed for human creativity to flower. At worst, politics and polititians may be truly decisive inhibitors. Nevertheless, one way or another, politicians have taken responsibility for development.

The phenomenon of public sector efforts, as noted, is quite new and a brief mention of its size and scale of effort is in order. In a fascinating essay, two young writers recently put the public sector in its contemporary dimensions.[1] The essay primarily dealt with inflation, but inflation and growth of the public sector are interrelated. In their essay, the authors chose to use the word diffusion rather than inflation, as they talked of the costs of this new phenomenon, the public sector.

These costs have to be paid for, and such payment can be achieved only one way: through taxes. But it is more subtle than mere imposition of a direct levy. The taxes we now talk of, this enormous preemption of funds to pay for the public sector, is not just a direct charge levied by the government on an individual, rather it is a charge on everyone. Taxes are diffused throughout the economy to pay for the public sector, and the consumer is the payee, with the poor paying most of all. Thus, when taxes go up to pay public sector costs, so does the price everyone pays for all goods. The cost of the public sector is diffused. This is the first fundamental we must accept when we approach government in action.

Thus, while public sector assumes responsibilities for agricultural development, especially in the Third World, we must realize that the cost of this development will be diffused and borne by most everyone. In the Third World, that means the peasants, who by their efforts produce wealth. But we must go further than this. Not only will development costs be diffused; so will the cost of *all* of the public sector. Military, political bureaucratic salaries and related expenses; public buildings and limousines; government projects losing money; delegations abroad; and those almost unclassifiable public schemes and programs — it all has to be paid for by the citizenry, which mainly means the poor because these are in the majority. One is forced to repeat the truism that government is not "free," but has to be paid for by the people.

Let us assume the popular view that political systems are necessary. We can certainly further assume that they are here to stay and that their size and power will not diminish. Let us assume also that it is possible for political systems to work in the interests of people within the system's jurisdiction. This obliges us if we are to make a complete analysis, to turn the theory around and look at baser aspects of political systems, especially in action.

One can discern at least six basic features of politics in action, features of which any person to be affected by such action ought to be aware. First is a strong tendency to make promises which cannot be kept. This seems to be as much a part of democratic systems as of totalitarian. This tendency feeds upon itself. The politician makes a promise to gain favor with a constituency. Constituents more or less demand that promises be made. It is hard to decide which of these comes first, and it does not matter. The constituents have a difficulty and they seek remedies. They hear political claims for solutions to many things. A kind of metamorphism or exchange occurs. Politicians end up by making promises people want to hear. Exaggeration tends to be a by-product.

In democracies in particular, politicians are beholden to mass voters. It must be extremely difficult to stand in front a large audience and say government can do only a little on behalf of individuals and that most sucesses in one's daily life will arise not because of government action, but because of individual action. It must give politicians a momentary, if not lasting, elation to recognize a public trauma and respond to it with the right kind of promise. No doubt, too, there are politicians who make profit from the psychic value of promises.

People want bread or sustenance. They are hungry. The politician promises people bread. It is, he admits, their bread, made by the people's own hands. He takes it from them, he admits, and gives it back to them from his hands and they are grateful to *him*. For this, they give him their allegiance and follow him happily.

Allied to promise is the second factor, a corollary. Promises often lead to lying. Lying has become a common feature of contemporary politics. Adolf Hitler used it quite conspicuously. If the lie were big enough and told often enough, it became "truth." Orwell was contemporaneous with Hitler and savagely indicted the same technique both in *Animal Farm* and *1984*. Language became inverted, "peace" meant "war," and "right" meant "wrong." Orwell wrote of his times. Later Churchill observed that if a political leader were faced with a position where national security or telling the truth were the issue, then the leader must lie.

This becomes especially true in contemporary politics where increasingly government is becoming involved in practical things, things which are part of reality. Rhetorically it may not matter too much in practical terms that the condition of the country, described as "impoverished" by a political aspirant, becomes "prosperous" after he is elected. This tendency of politicians matters a great deal, however, if a political leader decides to become directly involved in the working of a nation's economy. Because of his looseness with terms, he has posited two opposed views of a reality, because the economy has not changed with his coming to power. It is merely his status that has changed. Here lying can become quite serious in that it probably misleads the politician as much as it does his listeners.

A third characteristic common to government is a lack of specific knowledge as to how things work. When government was less actively involved, this did not matter too much. Now governments are involved, and with complex activities. When we look at complexity in contemporary society we are inclined to focus upon the intricacies of nuclear science or the ramification of some new scenario such as Law of the Sea. But the everyday working of an integrated economy is complex, perhaps so complex that no single human understands the full extent of it. It works because millions of people do what they know how to do. In a miraculous way it usually all comes together.

It seems presumptuous for any centralized body to suppose that it has knowledge to direct an economy, let alone to "fine tune" it — to borrow a contemporary, if now discarded, term. Yet this is a feature of modern government, we are told — this managing of such complexities.

Forturnately this management is often more rhetorical than real, even in those states where government asserts its omnipotence. Fortunately, ordinary people go about their daily work happily ignorant of what the managers are doing. Nevertheless, interventions occur. Thus we must expect those jolts and shocks emanating from the political system as features of modern life.

This ignorance of the way things work also makes governments very vulnerable. This vulnerability can lead them to excesses as they protect their inefficiency, especially in states with totalitarian overtones. A further factor compounds governmental ignorance, especially in the Third World, as regards agriculture: governments tend to be city-based and more detached from agriculture than they are from other complexities.

A fourth characteristic, closely allied to the preceding one, is the tendency of governments to stumble into wars that they cannot handle. We have already noted international tensions as an inhibitor to agriculture. Often wars stimulate agriculture, as World War II affected farming in North America, Australia, New Zealand, and South Africa. These were secure areas. In other enormous territories, China, Japan, Europe as far as the Urals, between 1939-45, war was a devastator.

One is struck by two phenomena regarding war, its frequency and the trivial reasons for it. From the Peloponnesian Wars in the 4th Century B.C., to the present time — a period of about 2,300 years — Western Europe has had less than three centuries of peace. Perhaps it is unfair to blame politicians alone for such a terrible record. We can instead blame human nature or intangible forces in international relations. Nevertheless, one is left with sneaking doubts about government.

One cannot read the statements of Gibbon's Senators, if not Emperors, without becoming entirely convinced that it was a narrowly focused ignorance which stumbled these men from disaster to disaster. Power infected them as it still infects men, even though Gibbon's Rome has long gone. There is a Buddhist adage from the Dhammapada:

Without fear, go.
Meditate.

Live purely.
Be quiet.
Do your work, with mastery.

Politicians almost reverse this dictum — they fear, act precipitously, are corrupt and venal, and show enormous ignorance of reality. Is there any wonder they stumble into war?

A fifth feature of government has already been alluded to. Governments inflate the economy, debase the currency, create diffusion of costs often created by them. The public sector, which is another name for government, is now real — a feature of our daily lives. It has to be supported. Such support has to come from those who produce wealth.

Obviously, wealthy countries can afford bigger and more expensive public sectors than poorer countries. It would be fair to say, nevertheless, that everywhere the public sector presses against the productive sector. Governments do not produce material wealth. People working with their hands and minds do. Governments, however, have now decided to dispense wealth which they have not created and which they must first take from the people.

We, the people, have lost sight of this fact. We debate with one another as to who should receive low cost food in the U.S. through this or that program. We forget that before one can start distributing food to needy persons, food must first be produced. A creative act must have taken place arising out of efforts by individuals.

Much the same occurs when governments distribute other forms of largesse, including income. It would seem that this action by government can be borne only in degree. When government action inhibits the creative act, it inhibits the growth of wealth. When government seizes the power to dispense wealth it has not produced, as a rule it is transferring wealth away from a productive sector into a non-productive sector. This becomes a direct cost which has to be borne by the productive sector. The two actions interact. Call it diffusion or inflation; it does not matter. The result is much the same, a debasement of purchasing power and a diminution of material living standards.

Sometimes it seems that we forget the real cost of inflation or diffusion of nonproductive cost. We are inclined to measure inflation as an economic abstract. We pride ourselves that inflation is only 6% here while it is 12% over there. The only reason for pride regarding inflation is when there is no inflation at all.

The real cost of inflation is its impact on people. It is the elderly poor who are hurt most and first. Without capacity to increase their incomes, they can do little except endure, to a bitter end, a diminishing material welfare. Next, there is less wealth to be spent on art, museums, libraries, churches, the theater. Education must be curtailed because it subsists on surpluses arising from productivity and these are now diminished. The last to be hurt are the young; healthy producers who can work harder and longer as they emulate the squirrel on the treadmill. Or are these the last to be hurt? Probably not — the last are likely to be politicians.

The sixth and last characteristic of politicians or government concerns power. We have been warned often about power and politics. Acton gave his famous dictum, and the very founding of the American Republic was about power. The Founders recognized the nature of power and politics as did Acton later. Their entire thrust was: how was power to be limited and controlled? Power, its accumulation and its abuse, is a desease of politics. Perhaps it is the very nature of politics itself. St. Paul probably put it most directly, most dramatically, and certainly most terrifyingly.

> **For we wrestle not against flesh and blood,**
> **But against principalities, against powers,**
> **Against the ruler of the darkness of this world,**
> **Against spiritual wickedness in high places.**

Not all leaders abuse power. History shows us, however, that most do. Abuse of power by political leaders occurs with a consistency in history not given to most other aspects of politics. It surely must be regarded as an occupational disease of politicians. There are consistent features, too, regarding this abuse of political power, which, in itself, might be defined as the bending of a mass of people to the will of a few. The man bent on attaining power does not listen to the adverse. There is no need to listen, for unless the words are supportive, power-bent politicians regard contrary views as diversions, if not irrelevancies. This situation can lead to severe national problems because realities tend to persist and have their way. It is no way to avoid potential disasters, pretending they do not exist.

Men abusing power are also given over to fixed constructs. They believe they see the way ahead clearly and they will not be deterred in its pursuit. This may be fine for an individual's ego. Such attitudes on the part of leaders may bear heavily on humble people. After all, it is they who have to pay, both economically and personally.

The power-centered man also despises the individual and his creativity unless these serve him. Indeed, the power-centered man sees as his true enemy the individual operating with a sense of freedom. He especially cannot abide the balanced rationalism of thinking persons as they perceive the world about them from their individual viewpoint and seek to operate effectively relative to their own individual capacities. Wrote Hayek:

> **The fundamental attitude of true individualism is one of humility towards the processes by which mankind has achieved things which have not been designed or understood by any individual and are indeed greater than individual minds.[3]**

If power is abused to its limits, this leads to two other interrelated features. One is the abuse of the individual person by all of those terrible actions so typical of the more recent history of our time. Allied with this is the rise of the centralized totalitarian state. In a recent commentary it was noted:

> **Nesbit's thesis... was that the emergence of the "centralized territorial state" was "the single most decisive influence upon western social organization." The history of the West since the end of the Middle Ages was the story of the decline of intermediate associations between the individual and the state. The weakening and dissolution of such ties as family, church, guild and neighborhood had not, as many had hoped liberated men. Instead it had produced alienation, isolation, spiritual desolation, and so, to satisfy his longings, he seeks out ersatz community — eventually finding it in the totalitarian state.[4]**

The end product of the abuse of political power, fortunately not always achieved, is indeed the totalitarian state. We must realize, however, that the totalitarian state is not imposed willy-nilly from above. Rather, it arises when much of what previously existed in the form of individualized institutions has been weakened or destroyed. The power-centered man in extremis happily aids, may even engineer such destruction in his total claim to power.

Politics must be viewed as a human phenomenon having many facets. At the center stand the models of the theorists and sometimes of founders of political systems. Here politics has a design and is institutionalized. This at least is the concept, but one wonders if this is ever entirely the case. In agriculture we have noted two systems where attempts to mold farmers and farming to fit a particular theory have resulted in agricultural mutation bought at the cost of enormous human suffering. The Soviet ideal of a collective farm, communally-owned operators with decision making, and with marketing and proceeds communally controlled, is a far cry from the state farm system. Here ownership is by the state. Operations and control of results emanating therefrom are those of a state bureaucracy and not of a commune. We can only reflect on the price of all this in human terms as well as the enormous vigor and vitality required to bring about this creation of yet another bureaucratic machine.

In China, that other notable example where politics was to determine agricultural method, we see an even greater departure from an original version. The Chinese have now turned toward a traditional form. This may be an interregnum. We suspect that Chinese traditionalism combined with Chinese common sense as this applies to productivity, will ultimately prevail.

In most other parts of the world, politics and politicians peck at agriculture. But waiting in the wings is this desire of a centralized power to have it their way, to bend farming to a particular political construct.

There is one further characteristic of modern politics — governmental indifference, even in the face of governmental desire to control. One often finds this paradox at work in the Third World. The peasant knows that it is from this seat of power that a vague authority emanates. In modern times this power is perceived by the presence of officials — police, military, increasingly controlled television, proclamations on internal actions and strong demonstrations of nationalism. The latter is expressed through such symbols as flags, publicized parades, and continual exaltation of the state, coupled with propaganda about threats from neighbors or sinister designs by a superpower.

In the guise of protecting them from these threats, government tends to control people. In some countries this may be reinforced by strong traditional carry-overs. In the past a centralized monarchy exerted authority in a similar manner. It expected peasants to render a labor service to the state, this being a strong traditional element in centralized control. It was, nevertheless, a loose control. It is no less loose today. Within this loose system, there often coexists a degree of indifference to real issues as these affect peasants.

Let us go now to a tiny village. It is the summer of 1972 and the story concerns a meeting between the village headman and a group of Western visitors. Questions were posed to ther headman about crops, markets, village social life, and village aspirations. Someone asked the headman about government, what did he know of it far away in the distant capital? Did government ever come to this village?

Yes, of course they came, the headman noted; not frequently, but quite often. These were local officials, but they always let him know that they represented the government in the capital. He stated that he was always very polite, so very polite to them. He did nothing to aggravate them for that might make them stay.

Questions from the Westerners persisted. Why, it was asked, do they come, if you usher them in and out so quickly, seeking to get rid of them? Did not these officials have some specific purpose in coming — to render you a service, give help? Did you, the headman, not tell them of the same village problems you outlined to our group? (The headman had talked of water difficulties, lack of adequate education for village children, and absence of any health services.)

The headman sat quietly on the floor. He looked at the alien group, not resentfully, nor in anticipation of any help from them, either. He responded that the visitors did not really understand the government. The government, he opined, was not interested in the same things that villagers were interested in. The government had other concerns. He beckoned toward a splendid frangipani tree in the courtyard. Some blossoms still remained, dark red, glowing against the green foliage.

"It is like that tree over there," the headman continued. "When it is covered with blossoms, it is very beautiful. Everyone admires it. We in the village are very happy when our tree has is blossoms. But the government, ah, when it comes to the village it only wants to see the blossoms. They expect me to show them blossoms, even when blossoms are not there. Then they are happy and they go away."

"I, however, am headman, and I must stay. I cannot go anywhere. I too must look at the tree and I do not always see blossoms. It is roots I must deal with. They are tough and twisted, but are part of the tree. This is what I am left with, those twisted old roots. I do my best."

There is yet another factor which detaches government from peasantry and gives rise to this indifference. It is a factor common to many countries of the Third World. Third World countries, especially smaller ones, tend to be city-states. One can observe these city-states, particularly in Africa, Southeast Asia, and South America. Characteristically these states involve a great mass of peasants, often comprising up to 80% or more of the population. There is a capital city. Other cities exist, but it is the capital that is the focal point, while peasantry provides the bulk of the population. But they live in the countryside, detached from the city.

All roads, physically and psychically, lead to the capital. It is in the city where one finds the concentration of those appurtenances which make up the modern state. Here is the seat of government. This is where the politicians are concentrated. Politicians in poor countries rarely find a rationale in the countryside for their existence. Power resides in the city. To obtain power means a career and possibly other opportunities. One does not become an ambassador abroad or lead delegations to international conferences by locating in some distant province. Within the system, top jobs at home are gained by men on the spot, not merely in the city, but right at the focal point of power.

As an aside, although Mao Tse-tung and his immediate entourage started working in the cities, they eventually found power in the countryside. In 1927 and 1928 they attempted to seize power in the cities, using the city-based proletariat. This was sound communist doctrine. Mao's forces were defeated, and nearly destroyed *in toto* by Chiang Kaishek's forces. Then Mao retreated to the countryside. As the son of a peasant, he was already familiar with the countryside and peasants. In the years that followed the Long March, Mao and his entourage became immersed in peasants and peasantry, and Mao's doctrines of communism changed. Military strategy changed as well, being based upon peasants. In one notable instance, years later, there was even a peasant-oriented strategy to conquer the world![5] Much the same occurred to communist leadership in South Vietnam, Cambodia, and Laos, during the war of 1961-74. Communist leaders were driven into the hinterland. This has occurred also in Thailand, and before that in the Philippines, Burma and Malaya. This aside is offered to throw into bold relief the almost total detachment of all the other city-bound governments from peasants. Further, in most peasant countries today not only politicians are city-oriented and city-based; the city also tends to be the home of the bureaucracy.

The tale of our village headman was not offered as a merely interesting and exceptional case of peasant life and of the poetry in a humble man in a remote hamlet. One the contrary, perhaps without the poetry, the condition outlined is typical. The "city-state" consists of officials, who do not know their own countryside because they are basically cut off from it. Often political leaders want to know what "it is like out there." Many political leaders know more of Paris, London, and New York than they do about their own countryside. They have only the vaguest idea of the geography. The way peasants go about their daily work is as great a mystery to them as is village life, organization and aspirations. It is also common to find that the city-based bureaucracy is ignorant of what it is like in the countryside.

There is, however, a distinct difference between bureaucrats and politicians. The politicians' ignorance is born of detachment: simply that most everything connected with power, career and tenure is located in the city and has little to do with the country. This detachment also applies to bureaucrats. But with them, another factor also exists. The increasing specialization of bureaucrats limits still further their knowledge of peasantry. In bureaucracy, ignorance tends therefore to be institutionalized and perpetuated.

The city, usually one great city, is also home to the military, major entrepot for exports and imports, banking and other general commerce. The city is invariably the home of the media. As might be expected, little (if anything) about the countryside is ever printed in city-based media. Media communications with the hinterland are generally poor and the only rural news one sees is of some disaster — a flood, raid by bandits on a police outpost, plane crash in some remote province, events which fortunately are fairly rare. Foreign news, on the other hand, coming as it does over international electronic networks, is often quite comprehensive and up to date. Through eyes of the media, it is almost as though the city were linked to the larger world, but detached from its own countryside.

To city-dwellers themselves, the countryside exists as a hazy outline in the background. Country people come to the city to become house servants. They fulfill other roles such as driving cabs, collecting garbage, peddling food, shining shoes, working on roads and other construction, dockworking and begging. This tenuous link between town and country is probably more tenuous today than in the past. Fifty years ago, today's great Third World cities were country towns. Their life-style then had more commonality with the countryside than it does today. There is an enormous gap between a peasant milieu and the city of today with its large hotels, television, movies, hospitals, traffic jams, smog, foreigners, jet ports, nuclear research, modern plumbing, electricity running to waste, sports stadiums, paved roads, latest fashion apparel, beauty parlors, supermarkets and night clubs. The peasant milieu has not changed so very much. The cities have. And the gap increases.

Compounding the division is a new city-based phenomenon, the arising of what might be called a "petite intelligentsia." Increasingly in the Third World, there is a small but growing number of intellectuals. They can hardly be described as intellectual pioneers or producers of new intellectual works. In terms of writing, research or development, intellectual productivity in the Third World is low. Rather, these "petite intelligentsia" are products of Western, or in fewer instances, Soviet universities. Invariably one finds this minority in the cities. Like intellectuals most everywhere, these people tend in part to be alienated from their society. Often their work, usually in bureaucracy or politics is not rewarding. Often power is held by old-line bureaucrats, or the military, or a tough political leader; sometimes by a combination of these. Further, it is in these power centers that wealth resides. In a poor country, wealth is hard to conceal. Indeed it is usually more ostentatiously displayed in the Third World than in the West — large automobiles, grand houses, fine clothes, bodyguards, and a general sense of deference by the populace at large. Where the "petite intelligentsia" is part of this power group there is little tension. But when the petite intelligentsia is not so located it becomes resentful of the inner circle.

These partially alienated intellectuals are also different from intellectual leaders of the past, especially in Asia. In Asia such intellectual leaders as Tagore, Ghandi, Bhave, Sun Yat-sen, Mao Tse-tung, U Nu, and Lin Yutang, despite the Western overtones of some of them, were apt to link directly to their cultures. They were in a tradition linking back to such giants as the Buddha and Confucius, who were cultural conservators. The new intelligentsia seem to be increasingly alienated from their own roots, largely as the result of their Western education.

The basic reference of these city-based intellectuals is not too different from that of most university campuses in the West. One obtains the usual Keynesian economic references and Western administrative methodologies. One also hears a good deal about democracy or socialism, but these philosophies often seem to be less aimed at the general welfare than they are toward betterment for the intelligentsia.

In general the "petite intelligentsia" are welcome in Western circles. They lend an air of authenticity to these ongoing relations: the intelligentsia are native, therefore real. In fact, cut off as they are from it, they usually know little if anything about the mass of their own society. They look native, but do not think as natives. For the most part they are the only people with whom communication by the outsider is possible. This, if nothing else, keeps these intellectuals going.

Chinese leadership on the mainland recognized that even they faced this problem. Their reaction was to send young intellectuals to the countryside. On the other hand, it could be argued that these new intelligentsia are part of the modernizing process. This line of reasoning suggests that alienation is precisely what is needed, being part of the process of moving away from a traditional society into modernization.

Possibly there is a grain of truth in this assertion, although this writer must express the gravest of reservations. The "petite intelligentsia," as far as can be discerned, are for the most part interested in modernizing their own particular milieu, not that of peasants. As has been stressed, their relationship with the greater society, especially peasants, is rudimentary if it exists at all. Neither is Western knowledge of much specific advantage. It all has to be adapted to suit a local condition, to obtain that critical mix between particularly indigenous and essentially new which allows things to work. One sees little evidence that the "petite intelligentsia" can or will work in this role. Thus under conditions that exist in city-based political systems of many Third World countries, one can hardly see them playing any but rhetorical roles.

Here then is an outline of the new "city-state," a condition which characterizes so many Third World countries. Probably only India and People's Republic of China could claim to be at least partially free of this syndrome. It might also be observed that the Third World is not the only region affected. As the role of centralized government increases, the tendency for an administrative capital to dominate obviously grows. This tendency carries with it all of the isolationist characteristics already noted.

In the Third World there is one other characteristic of the city-state. The city has to be paid for by someone. Most of these cities have no city-based productive enterprises. At best, they are service centers. At worst they are nothing less than a cost which must be met elsewhere. The tax base of the Third World nations must bear cost of the city. Unless the nation has particular industries beyond agriculture, the cost must largely be met by the peasant, for he is the only major producer of wealth.

Peasants tend to be taxed by indirect means. Direct taxation of peasants, such as income tax, is rare indeed. The peasant pays. On the surface, he does not seem to know it, although he often suspects that he pays the bill. These suspicions are commonly heard in conversation with a peasant group. If a group of peasants is asked what they think of Premier X, they will usually reply that he is fine, a very good man in fact. If the matter is carried further, such as noting that newspapers have implied that he is corrupt, that he took money from the public purse, peasant response is usually an acceptance of the fact and a notation that it has happened before. When asked would it not be better then to get a new Premier, the response to that suggestion is almost invariably negative. Why? Well, answer the peasants, we know this one and we know he has already had his fill from us. If we get a new one, say these astute people, he is liable to have "empty pockets."

Perhaps one is being overly harsh on cities of the Third World, these focal points of political power, cut off from their own people by the city dweller's own contrivances, but one thinks not. It is striking that Asian communists in particular have recognized this problem of separation. Mao Tse-tung specifically asked how a government in a peasant country could function as a sponsoring agency for peasant development, if it was apart from the peasant. In the Chinese situation other factors were also at work. The countryside has offered to the city dweller a kind of Rousseauean or Tolstoyian purification rite. At least this is what Mao seemed to be saying. By working in the fields with one's hands and listening to peasant wisdom, it was supposed that city persons could be regenerated and could then re-focus their minds and energy on realities. Maybe Mao had a point. For our purposes we are looking at politics and its relationship to the growing of food and fiber. Let us sum up as far as we have gone.

In the end we are attempting to create an environment promoting development, and this is where politics has entered our scenario. The predominant aspect of this environment is liberty, for liberty presupposes that individuals may be free to exercise their talents on their own behalf. It is a condition which contemporary politicians seem to dislike most. They seem to resent aggrandizement of individuals outside politics. They appear less interested that all worthwhile inventions geared to increasing productivity have a societal impact. People who do create or invent are expected to be faceless while social good emanating from their talents is publicly exalted by politicians. Rightly or wrongly, humans do not react positively to this kind of restriction upon reward of effort. Similarly, creation or inventiveness suggests inequality of persons because some persons have more of a particular talent than others. A feature of contemporary society is a pretence toward equality. Where society seeks such equality, especially equality of result, it will obtain mediocrity.

Yet among literally billions of poor peasant farmers the fruits of invention, innovation, and creativity of system are desperately needed. Such attributes offer to those billions their only hope of escaping from a materially crippling traditional agriculture. The talent exists, but it must be stimulated and released. Instead, billions are increasingly required to look to a tiny handful of political leaders for this release; here lies the crux of the problem. It is simply not feasible that talents required in such diversity and in such scale could reside in this small coterie of people.

Instead, the requirement is that those with particular qualities of leadership use them not to dominate others and aggrandize their pockets and their egos, but in the service of others. The leader should be a servant. The concept is not new. It is merely forgotten. There will be little effective sponsorship of human development through political processes until the concept is revived.

One can indulge almost indefinitely in general reflections as to relationships between politics and agricultural development. Along with ideology, with which politics might also be equated, perhaps it is here that future agricultural development rises or falls. We noted in a previous chapter the workings of ideology on a grand scale in the Soviet Union and PRC. Now is pertinent to take a more microscopic view of what politics is doing. One can philosophize endlessly about politics and politicians, but it is their acts that really matter. Viewing agriculture, especially in the Third World, the impact of political policies measured in terms of increased productivity is not difficult to discern. Productivity is not only extremely low, but in most countries it has been static for so long that any movement is readily discernable. And the only movement possible is upward.

In late 1975, the Comptroller General of the U.S. sent a report to Congress on this matter.[7] The report concerned itself with politically induced factors inhibiting agriculture in forty-six countries, mainly those of the Third World, and involving some 1.7 billion persons.[8] The report begins:

> **Developing countries can increase their agricultural production and provide their people with urgently needed food if they provide their farmers with economic incentives and supporting services.**
>
> **However, these countries [i.e., the forty-six countries surveyed] have policies and institutional factors which act as disincentives to their farmers to expand agricultural production.**

The report then proceeds to summarize these disincentives:

> ***Low producer prices* [i.e., set artificially by government] discourage farmers from using more production methods or otherwise expanding production.**
> ***Export taxes* restrict production for export.**
> ***Monetary and trade policies* make food imports attractive and discriminate against food and agricultural exports.**
> ***Restrictions on moving food* from surplus to deficient areas discourage production in the producing areas.**
> ***Institutional credit* generally is not available to small farmers, producers for export are favored over producers for domestic consumption, and problems in obtaining agricultural credit force farmers to use more expensive forms of credit.**
> ***Extension services* are generally inadequate, do not reach small farmers, and are applied to export crops rather than domestic consumption crops.**
> ***Extreme disparities in farm sizes and forms of land tenure* deter increased production.**

All these factors have already been raised in this book. The important factor to be stressed here is that they have all been politically induced. This calls for further comment.

It might also be noted that according to a declaration from the U.N. Conference on Trade and Development, the Third World has called for a 25% share of the planet's industrial output by the year 2000. Their present share is just 7%. Admittedly the method by which this rather startling demand was to be met did not match the eloquence of the demand rhetoric. For example, multinational ventures using private business as the instrument were sharply criticised. The ever-present threat of nationalization was muted. Nevertheless, the prospect of nationalization hardly makes the course of further overseas investment attractive to any investor. Even less attractive was the vision of regulation of various kinds. This more or less assured that joint ventures would be unprofitable.

Increasingly, there is less and less incentive for private corporations to invest directly or indirectly in agriculture. The investment needed to support agriculture lies first in building appropriate machines, fertilizer plants, transportation and basic processing. Although in peasant societies a kind of infrastructure already exists, that is, human capital, there is simply too much uncertainty for investment to proceed to any significant degree. The primary responsibility for such uncertainty is political.

In some parts of the Third World simple but deadly terrorism endangers the physical person of foreign entrepreneurs. This creates a destructive climate of pessimism. The issue here is that if Third World countries cannot handle such basic enterprises as their own agriculture, it becomes exceedingly difficult for those same countries to attract other development. As we will note later, there is a relationship between the two. Suffice it at this stage to note the relationship and return again to politics in agriculture as the two currently relate in the 46 countries studied.

It must be stressed again that the solution to the food problem has to be solved by countries concerned. Alleviation, but not solution, can come from increasing the food exports of the agriculturally developed countries. The GAO report stresses the obvious, that without economic incentives the farmers in the Third World will not increase output. The critical factor, however, is that this lack of incentives (or one should say the prevalence of disincentives) is totally political. The GAO report makes a further breakdown by countries, giving nine "basic disincentive areas."[9]

	Countries
Government control of producer prices	**38**
Government control of consumer prices	**35**
Government procurement practices for food crops	**26**
Export taxes	**22**
Export control	**22**
Restrictions on credit and land tenure	**19**
Import subsidies	**17**
Restrictions on commodity movements within the country	**11**
Exchange rate controls	**6**

The report points out that in 1974 the World Bank alone provided $4.6 billion to the Third World's agricultural sector. Yet all donor nations and agencies have been reluctant to use their economic leverage to insist that recipient countries change their ways. As the report states,

> **One delegate suggested that the discussion focus on what donors can do to help the developing countries and not on what the developing countries can do to help themselves.[10]**

This extraordinary inversion of common sense was not disputed, although later in 1974 the U.S. Secretary of Agriculture did obliquely suggest that the U.S. as a donor nation "would not shirk (its) responsibility to press for progress on these critical fronts."

It will be suggested later in this book that donor nations go further that this and withhold aid unless disincentives be removed. The world cannot tolerate this kind of waste, where man-made disincentives at home have to be counterbalanced by capital disbursements from a donor abroad. This is not merely a waste of material resources. The disincentives referred to are disincentives to people: to peasants who provide the sweat, take the risks, and all for pittances.

The GAO report gives many and varied examples of disincentives at work. Each has its own rationale for a particular locale. There seems to be little doubt that critical similarities also exist throughout the countries surveyed. These are our next concern.

The first is interruption of the basic law of supply and demand through price fixing. This at once fails to provide an adequate economic incentive for a farmer to increase productivity. Of 38 countries where price controls are noted, the report also gives a detailed analysis on eight as typical examples. It shows an extraordinary commonality as far as price fixing is concerned. The detailed analysis covers Indonesia, Sri Lanka, India, Pakistan, Kenya, Tanzania, Peru, and Uruguay. In every instance, price controls are politically imposed. They have arisen from promises made to urban dwellers to keep the price of food low. In general terms the selling price received by the producer is fixed at below-world-market prices, the home price is still fixed at the lower value. The government then subsidizes the price of those imports to keep them at the adjusted price. As might be expected this disincentive results in shortages under quite ridiculous circumstances.

> **It is clear, however, from recent experiences that reluctance to raise food prices (encouraged in some cases by the ready availability of food aid) has contributed to food shortages in many countries, by providing producers with insufficient incentives to increase production. The resulting shortages have brought increases in food prices that are perhaps greater than those that would have been necessary to bring forth a sufficient increase in domestic production.[11]**

These pricing policies have many other dimensions. Almost inevitably a black market arises which nominally pushes the price of certain items not only higher, but usually well beyond world prices. It is another artificial creation. Smuggling also develops.

The idea of physical force, always in the background, in the end will probably become a feature of the process. Farmers must be coerced to sell their products to the government and the usual method is that they simply have no outlets available other than what are usually referred to as "non-institutional outlets," that is, the black market and smuggling. Already there are reports of soldiers being used in Indonesia to force farmers to sell to the government. In 1974, when farmers in India attempted to circumvent orders to sell grain to the government, the prime minister threatened to invoke the Maintenance of Internal Security Act.

Under the circumstances, farmers attempt to use other weapons. They turn away from crops that they must grow at an uneconomic price. They tend to hoard grain for their own use, to sell to their friends at higher prices or simply blackmarket the product. Most important, the farmer will not use fertilizer. To do so, increases farmer's cost per unit of product without any commensurate reward. The farmer is thus being asked to subsidize the politician's policies twice, once through the below-market price, and then by increasing yields through expensive fertilizers, a cost which is not reflected in selling price.

A further significant factor is that inherent ignorance of the political decision maker is enhanced by erroneous feedback from the nonmarket place. Obviously controls required to institute and maintain this wholly artificial system become wide ranging and complex and very difficult to understand. The report states *inter alia:*

> **The World Bank responded that, with the existing maze of controls, the price system is so distorted that it transmits only confused information to the government and uneconomical orders to the producer. Because of the complex network of price supports, the two-tiered exchange rate system, rationing, subsidies, taxes, and other controls, conventional economic relationships become all but meaningless. The Bank stated that an efficient allocation of resources within the agricultural sector is impossible with present price distortions.[12]**

The quote given referred specifically to Sri Lanka. In that country one "free" pound of rice per week is offered to everyone except taxpayers (that is, payers of direct taxes), who form less than 0.5% of the populace and non-Singhalese tea plantation workers. Even rice producers and other farmers get their free pound. And "rationed quantities of rice, wheat flour, and sugar are provided at nominal prices to everyone." Sri Lanka offers perhaps the most extraordinary example of a nation with high food potential, yet with a serious food shortage. Farmers have simply been dissuaded from realizing the land's potential. To an appreciable degree, the Sri Lanka case is the story of all thirty-eight countries where politicians have decreed price controls on food. It is nothing less than simple exploitation of farmers to subsidize ephemeral industrial development (which does not occur). More importantly, it allows politicians to buy and maintain power through the urban population — that is, the political part of the population.

It might also be noted that Western food aid, especially under U.S. Public Law 480, subsidizes the posture further. But one should not overstress this feature. As a percentage of national food production in the Third World, grants from outside have been relatively small. However, there have been exceptions. The U.S. provided Indonesia with 46% of its cereal imports in 1970, 74% in 1971, and 33% in 1972. Nevertheless, this amounts to less than 5% of Indonesian cereal consumption for those years. Between 1957 and 1971, food grains supplied to India under PL480 totalled 59 million tons, or three-quarters of total Indian grain imports. Wheat is Pakistan's staple food. In 1971 wheat exported to Pakistan under PL480 constituted 87% of imports; in 1972, 76%; in 1973, 88%. Yet, in 1974:

> **The Pakistani farmer was paid $69.13 a metric ton for his wheat. Wheat prices in neighboring India were twice that amount and the market price was about $157 (f.o.b. U.S. Gulf ports, June '74).[13]**

It might also be noted that there is general agreement among agriculturists familiar with the area that Pakistan could readily be self-sufficient in wheat. But first, the Pakistani farmer would have to be freed.

Let us now view potatoes in Peru:

> **Peru has followed a policy of maintaining fixed prices. The government-set prices for potatoes, a primary food commodity, remained unchanged for extended periods while the cost of inputs increased sharply (for example, pesticide and fertilizer costs doubled in a single year). Apparently government prices did not cover producer costs and farmers stopped planting potatoes for market. Potatoes have been in short supply and have reportedly been sold at some retail outlets at black market prices triple the controlled retail prices.[14]**

Peru did not receive PL480 aid in potatoes. But the question can be raised, as it can regarding Pakistan and its wheat, and indeed for every other price-controlled foodstuff wherever PL480 is in vogue: should aid be given under these circumstances? Such aid in the end helps politicians retain control over foodstuffs, while at the same time inhibiting the country's farmers from increasing production. Aid under these circumstances is not helping production. It is aiding the disincentives to production.

The report again summarized the impact of controlled prices and food aid on the world food crisis.

> **Leading world authorities now indicate that such food assistance by the United States and other countries has hindered developing countries in expanding their food production and has thus contributed to the world food situation....**
>
> **A 1970 FAO publication stated, 'that it is clear...from recent experience that reluctance to raise food prices (encouraged in some cases by the ready availability of food aid) has contributed to food shortages in many countries, by providing producers with insufficient incentives to increase production. The resulting shortages have brought increases in food prices that are perhaps greater than those that would have been necessary to bring forth a sufficient increase in domestic production.'[15]**

Beyond pricing as introduced by politicians, the second greatest dissuader is agricultural taxes. Some of these are taxes on land area; land value taxes; net income taxes, marketing taxes, export taxes; special assessments and taxes imposed by marketing boards. Export taxes are the most popular with governments "because they are administratively easy to collect, especially where small producers predominate in the agricultural sector." There is an even more pertinent reason why export taxes are popular — the small farmer does not see this tax. Its inhibiting impact percolates down to him indirectly and he is only vaguely aware, if at all, that he is paying.

Of the forty-six countries involved in this GAO survey, I estimate that somewhere between 300 and 400 million farm folk are paying indirect export taxes. We should be aware that indirect taxes are not confined to Third World countries. They also are increasingly popular among governments in the developed world. Again the report sums up the impact of export taxes:

> **Such taxes increase the price of exported commodities, which decreases the amount a buyer can purchase with a given sum of money. A commodity in short supply is not seriously affected, but when a surplus exists on the world market, prices inflated by high export taxes cause buyers to seek other sources (1) resulting in a loss of needed foreign exchange; (2) influencing producers' expectations about future demand (these farmers grow less than they otherwise would).[16]**

In Pakistan, as one example, agricultural exports provide 80% of the foreign exchange. It has been noted that during a period of rising international cotton prices, the government nationalized cotton exports and raised export duties to absorb the difference between farm prices and international prices, rather than pass increased profits to farmers and middlemen. Later, when prices fell, the government did not adjust its duties downward.

Not all food producing countries export, so export taxes are far from universal. Nevertheless, the state will take its dues in taxes of different forms. In all 46 countries surveyed, with the possible exception of Uruguay, the various taxes levied on agriculture are punitive or nearly so. Their objective seems to be two-fold. First, it is allegedly to provide revenue for industrialization. Measured by the scale of industrialization in most countries, investment has been meager in the extreme. The second purpose of taxation is more directly discernable. It is simply peasants are paying for government. It is ironical that the government causes most of the peasant problems, if it does not also create a deep sense of injustice. Another aspect of government which these taxes pay for is arms. It is not uncommon for a Third World nation to expend one-third or more of its revenues on arms. Often all of this comes from general revenues.

Perhaps the saddest feature of the tax situation is that tax policies can, if properly constructed, stimulate agricultural productivity. Comparisons are odious, but we have already noted the impact on agriculture of U.S. tax laws. Where a tax system exists, a concomitant opportunity also exists to differentiate taxes so that productivity and efficiency are rewarded. This has been attempted in the U.S. Nor are productivity and efficiency the only assets obtained. It is likely that waste will also be reduced and conservation enhanced.

As one example from U.S. experience, pivot point irrigation systems, which produce those enormous circles so visible from the air, make the point. These great above-ground revolving arms spraying water and fertilizer on crops are much less costly to install than is the more common flood irrigation system. In the latter, water is distributed in canals and ditches and siphoned onto fields.

Further, under particular conditions, the pivot may use up to one-third less water than the flood system. The pivot obtains for its owner certain tax advantages. Currently, it qualifies for a tax investment credit. Its depreciation, assigned at a fixed rate, offers a further reduction in tax. Costs can be related to farm bookkeeping to save further taxes. In conservation terms, however, not only might a pivot conserve water, but like any irrigation system, it can conserve the soil's capacity to nurture plant life. The ravages of a drought disappear. Thus taxes can often offer incentives with a multiplier effect.

Throughout much of the world, and not only the Third World, taxes have the opposite impact. At best they become a straight financial burden. At worst, they diminish productivity itself.

The third disincentive mentioned in the report is credit. Rural credit is the lifeblood of any agricultural venture. Most agriculture is seasonal, with returns coming to farmers perhaps as infrequently as once a year. Seeds and fertilizers have to be bought *before* the crop is grown and marketed. The same conditions apply to equipment. It is simply impossible to consider agricultural development on any scale without rural credit. There is little argument regarding this point. Yet, at the same time rural credit, or its absence, remains a scandal throughout the developing world. The report gives random examples.

> **In Pakistan the 60% of the farmers owning the smallest farms got only 3% of the institutional credit.**
>
> **In Bangladesh, only a few farmers hold more than 3 acres, but these farmers received more than 80% of the loans from the Agricultural Bank and the Cooperative Banking System.**
>
> **In the Philippines the 27% of the farmers owning the largest farms obtained 98% of the institutional credit.**
>
> **In Thailand those receiving institutional credit held on the average 60% more land than the average farmer.**
>
> **In Tunisia, 90% of the farmers could not qualify for institutional credit.**
>
> **In Brazil, 3% of the farmers got 34% of the loans.**[17]

But these neglected small farmers do get financial credit. Besides institutionalized services noted above there also operates what the report delicately calls "non-institutional credit." Primarily this source comprises individual money lenders. One sees them at work all over the Third World. In fact, they often provide the only credit source for the small farmer, the man who needs credit most. These worthies provide yet another service, albeit a sporadic one. They are sources of information on demand; they are, in fact, informal marketing analysts. Often the money lender is remarkably correct in his analysis and up to date, if not prescient, with his news.

In the early 1960's, when floods nearly destroyed the jute fields of what was then East Pakistan (now Bangladesh), Chinese money lenders stimulated jute growing in northeastern Thailand. Jute farmers knew that the high prices might last for only a few years. Nevertheless, under the money lenders' stimulus and information, they took the risk. They changed from rice to jute. As it happened, before prices fell, private interests in Thailand built jute-weaving factories, which gave a certain stability to the growing of that crop in Thailand. Money lenders started it all.

The only problem with this vigorous and perceptive source of credit is the interest rate. It is usurious. The report claims that it can vary from 3 to 20 times higher than institutionalized credit. Thirty percent was usually the minimum going rate. Money lenders do, however, offer another advantage. Their service is prompt, flexible, and on the spot. Institutionalized borrowing, on the other hand, tends to be "rigid, cumbersome and time-consuming." As much as anything else, in borrowing from institutions, farmers complain about having to travel to some larger town where they feel strange and unwanted, and where officials often treat them badly.

What of politics and credit? Politicians claim the prerogatives of rural development. Thus the provision of credit, the very lifeblood of the ongoing agricultural process, let alone a developing process, has now become a political responsibility. Only in Japan and Taiwan is rural credit provided expeditiously and fairly to small farmers. In most every other situation, rural credit provided to farmers is a disgrace, a simple exploitation of the hardworking poor. Until this situation is rectified, there will be no major development, while at the same time the cost — the human cost — of rural credit exploitation will continue at its current high rate.

An interesting series of contrasts might also be noted at this juncture. Satellite pictures supplied by NASA show that the Texas-New Mexico border can be clearly defined by the agricultural differential between the two states. In Texas, agricultural development right up to the New Mexico border is well advanced. In New Mexico this is not the case, except close to the Texas border. Much the same situation, although not in such defined form, can also be seen between New Mexico and Arizona.

There is a clear reason for these distinctions, related to rural credit. Texas has very large in-state banks and can finance virtually any size agricultural operation. Although Arizona has a small population, it has adopted the branch banking system. Arizona's main banks are linked with larger out-of-state banks. Arizona, too, can finance large-scale agricultural operations. New Mexico, on the other hand, has small, individualized banks and limited branch banking. This has been a conscious political decision. It is more difficult, therefore, to finance major intra-state agricultural operations in New Mexico. Indeed, it seems that the New Mexico farmers along the Texas border who do have larger operations bank in Texas. The point is that agriculture is dependent upon credit, a free flowing revolving credit, at that. Deny credit though political or other action, and agriculture is denied. Not much happens.

Land tenure is another critical area covered in the GAO report. In many respects tenure is as important as rural credit and its implications are certainly more complex. Tenure is intrinsically linked, as one issue, with the control of farm profits. Involved is who gets what share for what degree of effort. Involved, also, is the critical issue of farm management, and not least, farmer morale. In the end it is farmer morale which really determines productivity. If the farmer feels assured that he is being fairly rewarded he will not only work effectively, but he will endeavor to work creatively, which is more important.

Land tenure is so complex that only the salient points, those related to productivity and politics, can be mentioned in this book. One should note as a first element that land distribution programs, so beloved of many politicians, are by themselves political cosmetics. They provide pleasing rhetoric and foolish promises; but seldom allow farmers to produce. For example, if pricing structures, taxes and rural credit situation remain unresponsive, while at the same time land is "redistributed" to peasants, the peasants are being fooled. At best they are merely changing one burden for another. At worst their new landholding can become a cost they simply cannot bear; or as in Communist China, they exchange landlord for government. Usually land redistribution programs under these circumstances decrease productivity along with concomitant diminution of the farmer's standard of living.

If politicians are to redistribute land by breaking up big estates to pass out smaller parcels to individual tenant farmers, the package must be comprehensive. It involves prices, taxes, credits and rural education. Two other factors need mentioning: Besides services needed by the new landholder, assuming the state is to be the provider, new landholder must be given a measure of freedom. It is freedom to make his own decisions regarding crops to be grown, where to allocate effort and resources, and how to sell. Thus the services the state offers, especially that of advice, must be in a delicate balance between suggesting and letting.

This is a rare combination, especially when politicians have made rash promises and see fulfillment of their own position and status in the accomplishment of some grand design. Here they tend to push and shove the peasant to do as they say. Freedom for individuals is not always pleasing to new political leaders. Many politicians feel that power, and the security of power, is having an obedient mass marching to the leader's drumbeat. Independent decision making is considered disorder, if not chaos, the antithesis of obedience and passivity. But if the farmer is to develop, he must have freedom in his new role. The only other way development might occur is through coercion of one kind or another. It is probably no accident that this is a most popular political mode.

There can also be another, more prevalent kind of freedom accorded the farmer on his new land. It is the indifference already noted. Once politicians have redistributed, they feel that they have done their job. The farmer should be truly grateful and everything should work for the best. The delicate balance of freedom coupled with a necessary service is neglected because no one really knows how to perform the service. Neglect is more common than anything else. Land has been redistributed with rhetorical flourishes. The new owners were formerly workers; they have little or no experience in management. They have no capital, and if they acquire some, it is at usurious rates. Their struggle becomes harder, not easier. Before, they shared half the crop with the landowner but had a little left over. Now they are on their own, bare before the political malfeasances noted in this chapter, while trying to become autonomous managers.

Another psychological state in land tenure is certainty, or uncertainty. Will redistribution stick? The state giveth; the state can take away. The report covers this point:

> **India's land reform program has discouraged the development of more efficient farming practices because farmers fear losing their land. The program provides that a family-holding not exceed eighteen acres of irrigated land capable of producing two crops a year, 27 acres of irrigated land capable of producing only one crop a year, or 54 acres of non-irrigated land. Some farmers are reportedly delaying the introduction of irrigation facilities to avoid losing part of their land.[18]**

Redistribution of land ought not to be taken as an equalizing process, yet this is often the political pursuit. Some farmers are more skillful, more hardworking than others. From year to year, this or that farmer gets a better break regarding natural conditions. Some farmers spend more; some save more. Some reinvest; others do not. If land is being distributed in a quest for " equality," then obviously redistribution will have to continue as the redistributor seeks to level out inequalities of effort and ability by the chimerical equality of compensating acreage.

Thus uncertainty arises. With such uncertainty related to volatile redistribution formulae, the farmer's goal will not necessarily be increased production. It will be to take steps necessary to hold on to his land. This could be the curtailment of production. Without doubt, the worst of the political ills surrounding land tenure changes have been these non-material factors, which in turn have a distinct material impact. That is the point of this book. Curtailment of the farmer's liberty, indifference to his situation after redistribution, or instilling of uncertainty as to his future tenure — these have been consequences of land redistribution programs in the Third World.

The report also noted politically-based restrictions placed upon intrastate transportation of food. By controlling distribution, governments (and the report noted eleven of them) seek to control prices. The policy is to provide cheap food for urban areas. The entire process is a form of control over people, both urban and rural. The report noted that in Indonesia, a land of 13,667 islands, 780,000 square miles of land mass, stretching for 3,400 miles or more east and west, the government restricted the passage of rice from surplus to deficit areas by control of inter-island shipments. Instead, the government sold these deficit areas imported rice (one million tons per year) at prices below import cost. The object was to keep prices low everywhere. The state competed with its peasants, using public funds as the leverage factor — funds raised in the first instance mainly from those self-same peasants.

India restricted grain sales between states to keep grain prices low to expedite government procurement. One consequence was that farmers in surplus areas had limited incentives to increase production. There is yet another consequence. "India also is reported to sometimes seek international relief for starving areas while some districts maintain stocks of surplus food grain."[19] Earlier, we noted some advantages of a continental system where agricultural surpluses in one area would tend to counterbalance deficits elsewhere. Yet we see the very opposite of a continental system which is senseless except as a means whereby politicians buy favor from urban people at farmer expense. Even here it is a short term program because curbs on productivity inexorably destabilize any economic system. In the end everyone has to pay.

The last important exploitative act by politicians against their farmers concerns control of exchange rates. These actions attempt to make food imports cheap by discriminating against agricultural exports that earn foreign exchange. This practice is a form of agricultural taxation — another tax which bears most heavily on the small farmer.

> **When a country maintains a single exchange rate that overvalues domestic currency, the exporter is "taxed" in that he receives less local currency than if the rate were more realistic, while importers are "subsidized" by being able to purchase foreign goods below their "real" value.[20]**

No doubt exchange rates can be juggled almost indefinitely. But the juggler must also face the fact that sooner or later it all catches up. It is a matter of gaining political advantage today to create greater problems tomorrow. One simple point: again it is the producer who pays. Once more we must add another disincentive born of political action to agricultural productivity.

Perhaps the entire point of this assay into politics is, in the end, to ask what is the political impact on human creativity — on human capital. One can hardly claim that the record is good, but it *is* edifying. The common trend today is for politicians to proclaim themselves as benefactors of mankind. They have declared, no less in the first and second world as well as in the Third World, their claim to be "fixers," men who make things work. They can create, so they say, the material conditions of the good life. Even more than that, they promise to create not only the material conditions, but the ethical conditions as well. Indeed, the ideologues among them proclaim that, through their formulae, they will actually create "new men." Other politicians have been somewhat more restrained in their claims, but the majority of these embrace the idea that they, too, are architects of human welfare, if not morality. Regarding this last category, the late Arnold Toynbee, in the penultimate chapter of his last book, took a middle view, although even this was a change of viewpoint for him:

> **This recognition that governments have a duty to provide for their subjects' welfare was a beneficent ethical advance in the field of politics. The State has now become a welfare organization, besides continuing to be a law enforcing and a war making organization, in most of the world's industrialized countries....**
>
> **...the objection is that the welfare state demoralizes its beneficiaries, and by the 1970's experience has shown that the specious objection has been partly borne out by events. In some of the countries in which public provision for welfare has been carried far, the feeling that it is a man's duty to earn his own living has declined, and — more disconcertingly still — a rise in the standard of living has been accompanied by a decline in the standard of honesty.[21]**

So whether it be in terms of rigid ideology or in some loose form of human reconstruction, one does not observe those remarkable changes in human character that were supposed to emanate from various political machinations. Even where the process of producing "changed people" was attempted by force, and indeed with great savagery as in the Soviet Union and mainland China, "new men" have not eventuated. It might be concluded that on this grand scale, political action has not had conspicuous, or even noticeable, success.

Rather, the alternate claim would be more true. In its grand design, and even in its smaller design, much political action has been exploitative and destructive. This has been especially true of agricultural symbiosis. We have noted in the case of 46 countries the nature of the political intrusion into agriculture. Prior to that we noted the relationship of ideology to agriculture, primarily in the Soviet Union and Communist China. In no country, at any time, could one fairly say that, measured by results — let alone results measured against human and material costs — mankind is assured by political and ideological leadership that the human race is moving to a better end.

Yet at the same time it must be accorded that politics and politicians now have a role, albeit a self-assigned one. Just as politics and politicians can create environments adverse to development of agricultural symbiosis, so might they, in small degree, create favorable environments. What we seek are those political constructs that help. There are few historical analogies to help us, but we might instructively look at one.

The Homestead Act of 1862 was a political act. This act was preceded by other action in the U.S., the aim being land distribution to individuals. These early actions had two origins. One: the Federal government needed money. Sale of public lands was one source for it. The other was a belief with deeper roots — that public domain belonged to the people. The head of each family was entitled to a home or a farm. Not only that, but possession of such should be in clear title and inviolable to seizure or foreclosure by the state.

In 1862 the Homestead Act was formally adopted. Any citizen or intending citizen, if 21 years old, head of a family or a veteran, could on payment of $10, file claim to 160 acres. Like many political acts, it did not fully accomplish its purpose of distributing land to individuals. Much land was already in private hands. Initially at least the state offered little support in the form of agricultural education for people who often knew little about farming.

Despite restriction, there was speculation on land. In fact by 1890, only one person in three retained his original possession. By the beginning of the 20th century, only one farm in ten in the U.S. was a homestead free of debt considerations. But these did amount to about 400,000 homesteads totalling 55 million acres. Those who sought larger estates outside of the Act by and large succeeded on a far vaster scale than was the case with the homesteaders. We should note the purport of the Homestead Act. It sought to optimize individual opportunity and to give the farm family autonomy. It highlighted the dichotomy between political acts that seek to create an environment for individual action and those that do not, some of the latter being noted in the GAO report. All political acts are not necessarily negative.

Sometimes authority badly misuses the vitality of individuals. We have noted the near incredible vitality of Russian and Chinese peasants as they were asked to recover from man-made disaster. One feels this same vitality exists throughout the world. The elitist mold of centralized politics is reluctant, however, to permit its release. Agricultural development, which gives rise to an independently operating peasantry spread over vast regions, is often not a goal of contemporary politics. Instead the seeking is for an obedient mass. The contemporary politician has a vested interest in keeping it that way. What we are witnessing is an ancient struggle dating back to the Greek philosophers and the Roman state. Does the state exist for the individual, or the individual for the state? Throughout most of the planet, based upon empirical evidence, the modern question can be answered only one way — the individual is now subordinated to the state.

As demands for agricultural productivity grow greater, new tensions, as yet unknown, are likely to arise. This book asserts that high productivity in agriculture requires individuals to operate in an environment which releases their creative human capabilities. They simply must have liberty. Freedom is the key, if not the operative word. It seems that all too often the state has other plans for individuals. At one end of the spectrum one can reflect on the kind of life Liu Shao-ch'i foresaw for individuals in People's Republic of China as expressed in a widely distributed pamphlet:

> **He [the individual] will also be capable of being the most sincere, most candid, and happiest of men. Since he has no selfish desires and since he has nothing to conceal from the Party, 'there is nothing which he is afraid of telling others' as the Chinese saying goes. Apart from the interests of the Party and of the revolution, he has no personal losses or gains or other things to worry about....His work will be found to be in no way incompatible with the Party's interests....A communist party member should possess all the greatest and noblest virtues of mankind....Our ethics are great precisely because they are the ethics of communism and of the proletariat. Such ethics are not built upon the backward basis of safeguarding the rights of individuals or a small number of exploiters. They are built...(on) the ultimate emancipation of mankind as a whole, of saving the world from destruction and of building a happy and beautiful communist world.[22]**

Liu Shao-ch'i, who in the time before he was deposed, was a leading theoretician of Chinese communism, gives the collective view, the idea of subordination of the individual to a greater good. Assuming that there are Chinese farmers who go along with this concept, one can remark that Liu Shao-ch'i did not bring about any great transformation of China's agriculture. There are other agricultural systems which place the individual rather than a political party at the apex. These do very much better. In the Liu Shao-ch'i context, one wonders in practical terms where one goes from there. If individuals do not want to be "good communists," does one coerce them? When one coerces them, what happens to rice production?

What we are really probing is the issue of power. Power to control the daily lives of people and the way they think. Such an aspiration is endemic to politics to some degree. It is difficult indeed for a political leader to face a national problem that may be well known to the majority of the people, and in turn say, "there is really very little that I can do about this — perhaps nothing. In the end it is going to be your efforts, more than mine, which will make crops grow." By force of circumstance, the leader is trapped. Because of folklore that has been built up around politics, he is expected to have a solution and be able to express it. There is further entrapment. What he offers is not a solution, for a solution can come only from the people themselves. Thus, as the situation does not respond there is more rhetoric, more promises, more lies.

But the problems of practical farming persist. Things may even get worse as various schemata confuse, obfuscate, and aggravate realities. So the whole obfuscation escalates. To a large degree this has been the pattern in the forty-six countries cited in the GAO report. In certain instances this etiology can degenerate to where ever more stringent forms of coercion are used to attempt to obtain results. Again to generalize, in part this has been the pattern of India's history until recently.

One is left with the question of whether political leaders in the situation outlined are victims or perpetrators. The end result is the same: an accumulation of power in the hands of a few people, a centralized authority. Then what of power? Is it always abused once it is obtained? Does it always corrupt? At the very least, history suggests that we beware.

There is an irony about the whole business of politics and agriculture. There is a role for politics in agricultural development. The nation ought to make, simply has to make, some collective decisions. Not only that, but ordinary people still look to political leaders to personify decisions taken in the general interest. A political duty is involved. However, the duty can become the temptation. Only persons of great wisdom who truly see their leadership function is to serve, will draw back at the right time — after they have gone as far as they should.

A nation needs a concept of itself and a concept around which it can coalesce. The farmer needs to have a degree of security induced by knowledge that he is a worthy fellow performing a function for which he will receive a fair reward.

Neither should we expect too much of farmers. I do not yet know of any who want to perform like Liu Shao-ch'i's "good communist." Instead they are concerned with plebian work of the field and succor of their families. They opt less for the Liu Shao-ch'i type of stricture leading to their perfection than they do for an opportunity to farm better. This is an immediate concern with intensely practical implications for their general well-being. What the farmer really needs from politics is less of it. He knows that politicians do not grow rice, but that he does. From the politicians he wants sufficient help to let him do it better and he wants to avoid disincentives created by the politician that result in his being less productive.

FOOTNOTES

1 David Marsh and Lawrence Minard, "Inflation is Now Too Serious a Matter to Leave to the Economists," *Forbes,* Nov. 15, 1976, pp. 121-141.

2 *Ibid.*, p. 127.

3 Friedrich A. Hayek, *Individualism and Economic Order* (Chicago, 1948), p. 32.

4 George N. Nash, *The Conservative Intellectual Movement in America* (New York, Basic Books, Inc., 1976), p. 53.

5 Martin Ebon, *Lin Piao, The Life and Writing of China's New Ruler* (New York, Stein & Day, 1970), pp. 197-269.

6 I am indebted to Robert Greenleaf, for reviving the concept of leader as servant. See Robert Greenleaf, *Servant Leadership* (Ramsey, N.J., Paulist Press, 1977).

7 Comptroller General of the United States, "Disincentives to Agricultural Production in Developing Countries," Nov. 26, 1975. This report was an expansion of a previous paper by Abdullah A. Salet, Foreign Agricultural Service USDA, "Disincentive to Agricultural Production in Developing Countries; A Policy Survey Foreign Agriculture," Washington, D.C., March 1975.

8 In order listed in the report: Mexico, Dominican Republic, Trinidad and Tobago, Costa Rica, Guatemala, Belize, Honduras, Nicaragua, Panama, El Salvador, Argentina, Bolivia, Brazil, Chile, Colombia, Ecuador, Paraguay, Peru, Uruaguay, Venezuela, Angola, Ghana, Ivory Coast, Kenya, Liberia, Morocco, Nigeria, Senegal, Sierra Leone, Zaire, Bangladesh, Sri Lanka, India, Pakistan, Burma, Indonesia, Malaysia, Philippines, Thailand, Egypt, Greece, Iran, Jordan, Syria, Turkey, Spain.

9 *Ibid.*, p. 3.

10 At the March 1973 meeting of the Organization for Economic Cooperation and Development.

11 *Ibid.*, p. 7.

12 *Ibid.*, p. 51.

13 *Ibid.*, p. 61.

14 *Ibid.*, p. 8.

15 *Ibid.*, p. 25.

16 *Ibid.*, p. 13.

17 *Ibid.*, p. 18.

18 *Ibid.*, p. 24.

19 *Ibid.*, p. 16.

20 *Ibid.*, p. 15.

21 Arnold Toynbee, *Mankind and Mother Earth: A Narrative History of the World* (New York, Oxford University Press, 1976), p. 584.

22 Liu Shao-ch'i, *How to be a Good Communist*, pp. 32-34.

9 AN INHIBITOR: BUREAUCRACY

Bureaucracies are very old, although the term itself is of recent origin. For our purposes, that of relating bureaucratic function to agricultural development, we begin by taking a common, if not classic, view of bureaucracy in action. A bureaucracy exists to implement policy decisions made by politicians. Bureaucracies also coordinate those actions required to attain pre-set goals. This being so, when the political process has taken on the task of agricultural development, then bureaucracy *per se* becomes critical to implementation of that development. That is why this chapter will focus on bureaucracy.

As noted, bureaucracies are very old. The Chinese civil service system helped rule the Empire for approximately two millennia. Constituted as it was, the mandarinate was the operating arm of the Emperor, the person responsible for governing China. Adopting the benignity that seems to come with long historical perspectives, the Chinese bureaucracy functioned well.

Bureaucracy was an important element in sponsoring the remarkable industrial and agricultural development of Japan, after the 1868 Meiji Reformation. It still plays an important and particularly constructive role in contemporary Japan.

In modern times bureaucracy takes on a new significance. It is argued that the modern state has become so complex in its endeavors and so large in size that a large, centralized bureaucracy is inevitable. This thesis is not only something of a simplistic overview, but also perhaps a self-fulfilling prophecy. Be that as it may, the argument for the inevitability of large and powerful bureaucracies as part of the modern state is widely accepted without question.

But arguments *against* bureaucratic functions in practice are many and varied. Bureaucratic ineptitude is real, and produces anger in the breasts of many. It is also expensive. Allied with ineptitude, this produces even greater anger, which stirs both the humble and the mighty. But there is much more to bureaucracy than mere ineptitude and cost.

James Burnham in his pioneering work[1] enunciated another thesis, that bureaucracy was the "dominant" factor in politics. This meant that ultimately bureaucracy not only administered policy decisions, but was also the shaper, if not the maker of policy. Persuasive arguments can be made in support of this concept.

Max Weber, who in the late 19th and early 20th Centuries pioneered modern concepts of the bureaucratic function, did not go quite as far as Burnham. Perhaps Burnham, who came much later, had more to guide him. Nevertheless, Weber did see bureaucracy as the key to control in any modern state. As Weber saw it, the expertise needed to make the modern state function was so refined as to make it imperative that the state employed professional experts.

Weber also saw the potential problem that might arise from his own requirement. If experts are essential and are employed in key roles, who controls the expert, other than other experts? In an esoteric field, politicians are unlikely to possess this kind of competency. The question was not satisfactorily answered. Weber also saw, as Gandhi did later, "a hardness of heart in the educated." This was what the modern requirement for expertise needed.

> **In sum, the more these ideal types of administration and rule of law are the more fully realized the more completely [they] succeed in achieving the exclusion of love, hatred, and every purely personal, especially irrational and incalculable, feeling from the execution of official tasks.[2]**

There he was — the personification of today's caricature of bureaucracy as the unfeeling machine, an adjunct to the computer, remorselessly exercising expertise in disregard of people.

Nearly a century earlier, in *Philosophy of Right,* Hegel saw the problem of preventing this machine-like expert,

> **from acquiring the isolated position of an aristocracy and [from] using the education and skill as a means to an arbitrary tyranny.[3]**

Hegel's solution was that, from above, the Sovereign would exercise his power of control over the expert while other institutions would exercise a similar control from below. Hegel behaved very much like other theoreticians. When from its own logic one's main theory creates a seemingly insurmountable problem, one merely creates a new sub-theory to neutralize the problem.

Besides the issue of control, two other difficulties were foreseen. Occasionally the expert within the bureaucracy would possess important, if not critical, information which he could withhold at will. Such information would be his own creation. Secrecy would become a potential force, enhancing bureaucratic power about which the politician could do nothing.

Another foreseeable problem was allied to secrecy and the assumption of power in general. How could the bureaucrat be made accountable? To this day, most of these issues have not been satisfactorily resolved. In our investigations, we shall see that expertise and overweening knowledge are not the real problem. Admittedly, our field of concern is quite narrow. Nevertheless, it may be that the thesis enunciated by the theorists is entirely inverted. Instead, we shall have to ascribe to bureaucracies other, quite different, qualities as compared to those of an aristocracy born of its competence and expertise.

We are concerned with practical things, with farms and farmers. Our investigation requires us to probe relationships of bureaucracy to farms and farmers in the contemporary world. As before, the main concern will be with the Third World, which is the main locus of the agricultural problem. We shall find that bureaucracies have developed characteristics which do bear appreciably upon agriculture. These characteristics are quite different from most of those envisaged by theorists. The main issue, nevertheless, is still power — its use, misuse, and abuse. As this examination is made, we must avoid treating bureaucracy as though it were something apart from politics and politicians. This may have been the general idea in the sense of a theoretical construct for a bureaucracy. In many instances bureaucracy is indeed the working arm of the political body, reflecting political decisions as to how national policy is implemented. Neither does activity necessarily come only from the bureaucrats. In other instances, the role is reversed. It is bureaucracy that has become quasi-political, generating its own policies along with its own *modus operandi.* One theme, however, is common to each situation. That is the idea of the centralized state as the apex of power. This is the key factor in our study of bureaucracy and its relationship to farms and farmers — power and its exclusivity, the independence of power. The Third World farmer needs help, and we shall later define the kind of help needed. The farmer needs help to bridge the gap that lies between the minimal production levels of traditional agriculture and an immediate doubling of that production level (followed by further doubling) through more sophisticated production methods.

The task is not only to introduce new agricultural technology; the catalytic factor in the agricultural symbiosis is work by humans. In this context, work covers a wide variety of human endeavor. The most critical factor in providing work for the complex symbiosis is human attitude, a particular state of mind. This is engendered by a suitable environment. Indigenous bureaucracy by design and by accident is an important factor in building an appropriate environment for human activity. Indeed this might be said to be the role of bureaucracy as envisaged by Weber and others. Those bureaucracies concerned have thus taken on a formidable task.

In the Soviet Union, agriculture — far from being traditional — operates on state and collective farms controlled by a centralized bureaucracy. The situation is similar in PRC. Just how significant that control can be, is illustrated by the events referred to in Chapter Seven. In the drama, sweep and scope of these events, the bureaucratic arm of government played a key operational role. This fact of centralized arms of government planning, directing, and controlling agriculture and agricultural change raises the first significant issue confronting bureaucratic direction in agriculture. This is the issue of size and function.

The question is clear: are there operations in human affairs that are simply too big and complex to be conducted effectively under control of one central authority? We must add a second conjoined factor to this question: when operations themselves are complex in their nature, is not the size of what can be effectively controlled reduced by the degree of complexity?

Many years ago it was reported that the People's Liberation Army in China, during the first phases of the war against the Kuomintang, once mobilized more than a million peasants to build a road in southern China. The sheer logistical immensity of having one million people concurrently and coherently apply themselves to a single task, is quite amazing. There was a single goal, building a road. No doubt the tools, work, and, and materials were familiar to peasants. No doubt, too, the relatively simple continuum of repeating the construction process yard by yard did not call for any significant learning process. Nevertheless, to mobilize a million people quickly for a particular task is an impressive feat.

What if the task had been much more complex than road building? A million people could hardly have been mobilized to build communist China's first nuclear warhead, or to formulate and implement China's foreign policy. Complexity introduces a new dimension to size. Road-building would obviously be easier to control than these latter matters. In the latter two there would be a much greater variety of interrelating complexities, thus a greater over-all complexity, especially in control and direction.

Agriculture, even traditional agriculture, is a sequence and a grouping of complexities. These require day-to-day decision-making, often of an instantaneous kind. In farming, one is involved with living organisms related to changing weather. Weather and soil are related in yet another construct. The living organisms have their own biological dynamic and time scales, quite irreverent of human concerns. The weather is equally irreverent. The farmer is involved in a constantly changing web of wide-ranging interrelationships. This dynamic must be attended to, usually on a daily basis, through sensible actions arising from sensible decisions. Lacking this, agriculture will almost certainly fail. As a sponsor, bureaucracies face a quite different task, even conceptually. Complexities faced by the individual farmer are multiplied into an entirely new dimension by factors of sheer size and geographical extent. And in the Third World it is expected to change the original complexity of localized farming into an even more complex unitary or unified system.

When one flies over continental U.S., one obtains a better grasp of the possible dimensions of such a problem, even though U.S. agriculture is not yet a candidate for centralized bureaucratic control. First one notices the immensity of the farming scene, not merely in size, but in numbers. Below are many, many farmsteads, silos, stock barns, equipment sheds and equipment *per se*. One can count a dozen or more different crops in any single flight, and these can be broken down into varieties, all slightly different, within each crop region. As one over-flies, the weather not only changes, but varies in its interrelationship to regions throughout this vast land. Terrain, altitude, and soils differ, and so does crop reaction.

One notices animals as well. A dairy cow offers quite a different challenge from that of a beef animal, just as dairy breeds themselves differ. One handles a more temperamental Jersey diffently than a more placid Holstein. Sheep on the Colorado mountains need different treatment than those on California's lowlands. Hogs, those intriguing animals, require the kind of attention normally bestowed on family members, even in this modern age.

Then there is the changing pattern of seasons bringing preparation, planting, growing and harvesting. One sees the critical adjuncts. Vast irrigation systems where flood irrigation requires management different from that of the newer pivot point sprinkler system, those great arms that often sweep more than two hundred acre circles throughout the Great Plains. In the right season, even from great altitudes in flight, the knowledgeable watcher can see dust of preparing or harvesting combines. Above all as one moves over this panorama one is conscious of people, tens of thousands of farm people. Most have given their lives to it. They work, act, think, and re-think constantly.

A practical person must be overwhelmed by the ignorance and presumption of a small coterie of people calling themselves a bureaucracy in their proclamation that not only will they direct all this, they will do so in the name of higher efficiency and higher good. There seems to be a fallacy in this, if not a hubris. The hubris is rooted in the concept that the man in the bureaucratic center knows more than the man on the spot. The man in the center may indeed have knowledge that the man on the spot does not have, yet needs as an essential element for breaking out of traditional structures. The hubris arises not because the man in the center knows he possesses such knowledge, but because he knows knowledge gives him control over the man on the spot.

We shall shortly examine the manipulation of this as well as other factors accentuating the tendency toward bureaucratic hubris. The main fallacy is that direction can be centralized in a bureaucracy. This is to disregard the factors of size and complexity. Many will protest that it is not fair to take U.S. agriculture as an example, because it is not controlled by a centralized bureaucracy. U.S. agriculture is much larger and more diverse than agriculture in other countries. There are, however, less than two million farm owners in the U.S. and about another four or five million farm workers. In India there are probably one hundred million individual farms and India's agricultural symbiosis, though different in terms of work, is no less complex than that of the U.S.

In other regions, Africa and South America, for example, the situation is not too different in scale from that of India. The Chinese pattern would be about half again as large as India's. Yet each nation-state has its own bureaucracy and these often operate within a territory which is vast and complex. From an agricultural point of view, it is totally artificial to imagine that this vast size and complexity can be effectively controlled, and efficiency attained, through a centralized agency.

Nor is the problem of managing size and complexity solely that of government. It applies equally to large business ventures. They, too, face the same dilemma, although in this case the dilemma is more quickly discernible and is usually thrown in bolder relief. In a perceptive piece, the deputy editor of "The Economist" made the point:

> **During the Henry Ford manufacturing age about 40 of the world's 159 countries have grown rich because they were temporarily able to increase productivity by organizational action from the top: *i.e.*, executives sat at some level in the offices of hierarchically run corporations and arranged how those below them on the assembly line could most productively work with their hands.**
>
> **This method of growing rich has run into two rather fundamental difficulties: a 'people problem' because educated workers in rich countries do not like to be organized from the top; and an 'enterprise problem' because, now that much of manufacturing and most of the simple white collar tasks can be gradually automated so that more workers can become brain workers, it will be nonsense to sit in hierarchical offices trying to arrange what the workers in the offices below do with their imaginations.[4]**

While agriculture is not automated in the industrial sense and is unlikely to be in any foreseeable future, it does call for daily enterprise and imagination. Agriculture, even quite primitive agriculture, is highly individualized, parochial and intimate. One should not assume that farmers can be directed from the top any more than can today's workers in other industries. In both areas, agriculture and industry, and for not too different reasons, inherent problems arising from the management of size and complexity are similar. The private corporation, however, has a regulating device which bureaucracy as a manager of a size-complexity operation does not have.

> **The main reason why bureaucratic production can no longer work is that the decision-taker in any official production system must now restlessly ask, 'What is the best quickly-changing and labor-saving technology that I should use to accomplish this task?' In a state monopoly decision-blocking power quickly falls into the hands of people who can explain most suavely that any boat rocker who keeps asking these questions is being a bad colleague. In some big business corporations, with a layer of management sitting upon layer, decision-blocking power has fallen into the hands of similar middle bureaucracies. These will go bust.[5]**

A rider should be added to this last sentence. Some will "go bust," usually the smaller ones. If the corportion is big enough, it may be nationalized or even subsidized by government; thus its inefficiency becomes compounded and institutionalized. Nevertheless, the point remains. Large operations are difficult to manage when they approach a "critical mass" of large size and greatly complex. We do not know enough to define precisely either critical size or critical complexity. Obviously they exist. When the critical point is passed, a stress situation arises. If motive is to attain effective results at minimum costs, the situation is basically unmanageable by centralized authority. These factors of size and complexity give rise to a bureaucracy of increasing size and complexity, which again compounds the difficulty. Bureaucracy is usually associated with government. It takes on a series of distinct attributes, about which more later. Private organizations can also become bureaucratic and display characteristics similar to government bureaucracy. But in truly private organizations there is a self-regulating device which often works. Bureaucratized private organizations, if they must be self-sustaining (profit-making), soon cease to be self-sustaining and die. They fail to make a profit.

In looking at sponsorship of agricultural change in the Third World, the task is of immense size and complexity. This holds true even for quite small countries. For large nations such as mainland China, India, Pakistan, Indonesia, Bangladesh, Mexico, Nigeria — all countries with more than 50 million people, most of whom are small farmers — the size-complexity ratio in agriculture becomes a major historical human problem. If humanity is to face up to the task, it ought to realize that it faces perhaps its greatest challenge. And we expect bureaucracies to meet this challenge? Let us place the situation in a less uncomfortable dimension.

Viewed either globally or nationally, that great mass, the people, represents a tremendous resource. The tragedy is that the bureaucracy tends to neglect human capital completely as the major resource of agricultural operation. It holds to a fallacious dichotomy, that of governors and governed. The former is the repository of power, knowledge, techniques and resources. The latter is a recipient that, from an unleavened mass, is to be instructed and shaped into something new. It seems sometimes that this is precisely the trap we have fallen into. The dichotomy can also become so hardened that it produces a communication gap which is well-nigh unbridgeable. This construct of governors and governed — the former seemingly active, the latter seemingly passive — seems all too common. There are reasons for this situation which must be understood before any serious attempt can be made to make bureaucracy an agent of change.

The first reason is that city-based bureaucracy is almost invariably aloof from peasants and farming. Bureaucrats are part and parcel of our "*petit* intelligentsia," perhaps its most permanent component. All bureaucracies today tend to be self-sustaining, with their own rules and regulations, their own internal system of accountability, tenure systems, regularized promotions and impenetrable *impedimenta* to job dismissal.

In Third World countries there are other factors which accentuate this self-centeredness of bureaucracy and contribute to its city-based isolation. Bureaucrats in the Third World are a minority, educated beyond the masses. So much so, that an educated person must join the bureaucracy. Further, it is within bureaucracy that one finds the vast preponderance of people educated abroad. Even more than with politicians, these folk have greater affinity with their American or European fellow alumni than with those they administer.

It is a very exclusive group, an elite with tightly closed ranks. This exclusivity arises as much by accident as by design. Peasants, that great eighty percent mass which is being governed, simply do not go to college or university, to say nothing of higher education abroad. A peasant can barely obtain the requisite education to qualify for high school. There are exceptions, but few. Even in the exceptional case, when a peasant does qualify for bureaucracy, he tends to remain outside the inner circle. He usually becomes a minor official in some obscure province. It is even more unlikely for a peasant to study abroad, even at the non-college level. Competency in a foreign language is required. Peasants simply do not learn to speak foreign tongues. Further, social barriers against their joining bureaucracy are formidable.

The bureaucracy remains the preserve of a city-born, city-based elite. The whole situation engenders total isolation from the bulk of the population. Bureaucrats are isolated by birth and city upbringing; by their education, especially if abroad; by their parents and their wives (for city-based bureaucrats do not marry peasant girls); by the demands of career-building, and by the isolating luxuriousness of city life comforts. Isolation compounds isolation.

This makes it difficult to go to the countryside. One feels alien and so can remain almost totally ignorant of the hinterland and its people. These folk can be more ignorant than politicians. I remember once driving in an Asian city with a young American-educated bureaucrat. His chauffeur stopped behind a truck laden with hogs. "I am ashamed," said the young man. "Time and time again my Western friends see these hogs trucks and they must be appalled at the way we treat these animals. Our country folk are still barbarous."

The hogs, which incidentally were in prime condition, were loaded in Chinese style, which to an uninformed observer does seem strange. Each hog is encased in a sturdy, cylindrical wicker basket. The basket is extremely strong and spaces between the woven wicker strands allow for ample light and air. Each hog within its own little wicker cylinder is laid on its side, and the cylinders are stacked like cordwood on the truck bed. Today, in many parts of Asia, with large capacity trucks, it is common to see several hundred animals, many baskets high, so loaded.

The Chinese have been transporting hogs in this manner for centuries, perhaps millennia, and with good reason. Hogs are peculiar animals. If they are loaded free onto a truck, that is, loaded loose with feet on the floor, unless they are very tightly packed they will invariably press against each other usually in the direction of one corner. It is a strangely serious condition peculiar to these animals. The longer the journey, the more intensely they crush against one another, suffocating many animals. The reason Chinese pack their hogs in baskets in obvious; they wish to preserve a valuable animal.

This was explained to the young bureaucrat. He was surprised that his country folk could be so wise. This story is not at all apocryphal. It is merely one example of the consuming ignorance of city-based bureaucrats about their own country. This is an ignorance brought by isolation, physical isolation from peasant reality through class, education, aspiration and physical location in a city. There is, however, a second kind of isolation leading to another kind of ignorance, an institutionalized ignorance at that.

Institutionalized ignorance is common to all bureaucracies, probably more so in the West than in the Third World. It is born of the high degree of specialization of modern education, together with the fairly rigid process of streaming persons through the bureaucratic career system. Bureaucrats tend, more and more, to be specialists. They can become so specialized that they know little about the primary purposes of the parent department to which they belong. A specialist on a particular aspect of an animal disease can know next to nothing about agriculture, even though that may be his parent department. The career of this person can be maintained only by his staying with his specialty. To break out into plant diseases or even more adventurously into processing or marketing is simply not done. He could hardly be expected to compete there.

In any case, it is not the practice. Promotion has been institutionalized into a particular form by the bureaucratic system. One is promoted with his genre. This is unlikely to change. Change would mean restructuring of the bureaucratic system, less emphasis on specialization in education and more upon general structure, plus focusing upon working to objectives. The general objective here must be development of human capital. The highly specialized nature of bureaucracies, together with their patterns of promotions, tend to institutionalize a high degree of ignorance in terms of such an objective. Increasingly, bureaucracy becomes less attuned to human operations.

There are exceptions to bureaucratic ignorance. In fairness, mention must be made of some worthy folk. One sees these few individuals all over the world, almost invariably in remote places. Usually they have come from the city but have given themselves over to the country. As a rule they work in complete obscurity, almost as if their city-based superiors did not know they were there. They carry on the concept of agricultural extension work in the best sense. They are rarely idealists, men with a burning desire to save humanity. Rather they are men who are deeply interested in their work. They like what they do and they do it well. These are rare individuals. One sees them only infrequently. They are so few, they barely make a difference. But they do give one some hope. If a bureaucratic system, no matter how inept, can throw up such people, it ought to be able to throw up others — and more and more.

If ignorance born of a multifaceted isolation is one feature of bureaucracy, an allied condition is a natural antipathy to burgeoning and vigorous individual creativity. Perhaps in this instance, one should refrain from such a general term as human creativity and use a more specific term, entrepreneurship. Entrepreneurship has to be a feature of agricultural development. Indeed the whole purpose of the endeavor is to bring entrepreneurship to life. Further, it is entrepreneurship at the farm and not the bureaucratic office of which we speak. This is the nub.

> **The defining characteristic [of entrepreneurship] is simply the doing of new things, or the doing of things that are already being done in a new way (innovation).[6]**

This is precisely what traditional agriculture needs, and we are viewing the bureaucracy as one alleged stimulator of that need. It is the operating arm of a national policy which we assume will be to seek agricultural development.

Why should there be an antipathy between bureaucrats and entrepreneurs? The primary reason seems to center around the idea of centralized control, which seems inherent to bureaucracy. Centralized control presupposes a sense of order — each person in his place, fulfilling his function in accordance with preordained prescription. One often feels that a bureaucrat would be more at home in a classic medieval society. In such a situation not only is it difficult for innovations to arise, but it is almost impossible for innovators to flourish for any length of time unless they operate outside the law.

In practice, even if there is such a preordained prescription in a given country, it is almost certain that the prescription is not being fulfilled. Bureaucracy simply is not efficient enough to insure this, even in totalitarian states where force is either overtly or covertly used. Nevertheless the conflict remains. The innovator or entrepreneur has to fly in the face of the system because he is by definition refuting the system. We are now talking as if there were only one entrepreneur, whereas what is needed, especially in the Third World, are thousands of entrepreneurs.

The problem thus broadens and still another dimension must be added. The government, in this case represented by the bureaucracy, is city-based. Just as the bureaucrat may regard country people as remote, so too do country people regard government as remote — hard to see, hard to get to, somewhat aloof and mysterious. The people see government as remote, arbitrary, selfish, exploitative and harsh. These feelings may run deep at cultural roots. Ancient monarchies, feudalism, and colonialism may have left their mark. Even in its modern form and even when benign, government may have difficulty in sloughing off such a basic traditional viewpoint. A humble farmer is reluctant to flout tradition concerning something he barely knows, by acting in a highly individualistic and conspicuous way, by becoming an entrepreneur. Yet this is what development is about. It is as much a state of mind — individually and socially — as it is anything else.

Entrepreneurship also means competition between individuals. Here, too, the small farmer may be restricted by traditional modes. In traditional societies, it is not the custom for individuals to break out of the mold. The tendency is to hew to the cultural line. This is so very strong that it takes an equally strong individual to take off on an entrepreneurial venture. Entrepreneurship bespeaks a certain independence of mind, if not accumulation of wealth. A peasant's new-found wealth seems to bother bureaucrats enormously. If wealth becomes great enough, bureaucratic attitude often changes to alliance, even obeisance. At incipient stages of accumulating wealth there can be resentment, not merely over money, but over the new state of mind. From subservience and dependency, the peasant jumps toward self-actualization and independence. It makes bureaucracy seem to be unwanted and unnecessary (which indeed is the case in substantial degree). The overall bureaucratic climate for entrepreneurial venturing is not favorable. If bureaucracy is to act as a sponsor of agricultural development, the inhibition against entrepreneurship must be broken. Several things might be done in this regard.

First is to change the bureaucratic mind. This probably means a dramatic reorganization of bureaucracy itself. Put in bluntest terms, a developing country cannot afford a city-oriented, detached and ignorant bureaucracy, geared toward exalting the bureaucrat and antipathetic to peasant aggrandizement. The bureaucracy must instead become a field organization, concerned with instruction and service. It must leave regulation for later, if at all. Encouragement of individual peasant entrepreneurship should be one of the bureaucracy's primary goals. This posture should involve trying many ways of doing things simultaneously. Competition should be part and parcel of the *modus operandi* and there should be no penalizing of people as regards rewards.

The county fair is a simple example of this new state. The county fair is a marvelous institution. Here the individual can not only demonstrate his entrepreneurship, but be rewarded for it. Here other individuals are educated in the best possible way: by the demonstrated example of their peers. Unless one sees, one cannot conceive that such bulls exist, that such an udder could adorn a dairy cow, or that such skillful application of work could be made by machines.

More importantly, one learns that the individuals possessing these marvels are not rare creatures at all, but persons like oneself. An urge comes upon one. One sees not just that great udder, but a herdful of such udders; one's own herd. Why not? County fairs will not solve the entrepreneurship problem by themselves. However, they illustrate the state of mind needed by any sponsorship group...simple, honest and practical encouragement of individual effort directed toward wise and more productive ends. One cannot repeat too often that the key we seek is a release of human activity. Bureaucracies rarely see the world that way.

A second step in bureaucracy reform is to decentralize it. Bureaucrats should be dispersed into the field, where agricultural development has to take place. It does not occur in a city; it occurs on farms. Decentralization also would test the capacity of individual bureaucrats to act as teachers. It would test the validity of specializations. Above all it would test the validity of the entire sponsorship of the bureaucracy. Such testing might lead to development of new policies and practices more in tune with reality.

Bureaucracies show yet another tendency of often stultifying proportions. This is the tendency to regulate. It might well be that the real role of bureaucracy centers around this point. Is a bureaucracy's purpose to regulate, or to render service? If it is to regulate, then to regulate as to what end? To the preservation or enhancement of some common good, or merely to control the lives of individuals? The question is abstract, for nowhere is bureaucracy effective enough to control individual lives without actual coercion or the threat of it.

For example, in the U.S. which can now boast of a lightly regulated agriculture, the issue of service or regulation is well illustrated. If all the extant regulations in the U.S., directly or indirectly related to agriculture, were put into operation, then probably not a grain of wheat would be grown and neither would a steer be fattened. As it is, probably no single person even knows all the pertinent regulations. The flow of regulations has now reached such a pitch (an estimated 8,000 for 1975 from the agricultural bureaucracy over all) that no single person can keep up with the output, let alone know the entire accumulation. On both sides of the spectrum, bureaucrat and operator, there is a tacit understanding that most regulations can be ignored either partially or entirely.

Regulation presupposes that the man on the spot does *not* know best, but that the central authority despite its ignorance and isolation does. It presupposes that the man on the spot is at least venal, and possibly criminal — that he will do things in his own selfish interest inimical to the public good.

The first supposition, that the bureaucrat knows best, can easily be disposed of. He may know about a particular thing: about a chicken disease, a marketing trend, fertilizer supplies, a new development in irrigation, how to hybridize plants or the like. This does not by itself confer upon him the right to regulate a complete system such as agriculture.

Here we face an ancient human dilemma, the relationship between specialized knowledge and generalized policy-setting and direction. As far as farming is concerned, the two have a distinct relationship. Specialized knowledge is essential, but it must be placed in the whole, as part of the agricultural system. In terms of transactions with government, this calls for certain kinds of human relationships. The specialist and the farmer must relate to each other on a cooperative basis rather than as adversaries. The county agent system in the U.S., especially in former years, gives an example of such cooperation.

If we must have regulations, they should not be an arbitrary dictum, but a teaching process, in which both regulator and farmer learn and adjust. In particular, the specialist must have personal contact with farms and farmers. This interlinking has been almost entirely lost in most regions of the world. Specialized knowledge can be critical to enhancement of health and wealth. It does not by itself confer a right to change a system without first considering the prerogatives of that system as a whole. If this simple premise were accepted as part of a bureaucratic creed, much hostility and tension dividing bureaucrats and operators would be eased.

It is less easy to discern what the regulators foresee in terms of farmers and the environment. It is such an emotional issue with so much ignorance that rational analysis is difficult. Because few people seem quite certain that a serious situation exists, yet at the same time, no one knows quite what to do about what might exist, regulation seems in order. It may not make much sense, but it appeases passions.

Coyotes are without doubt the most intelligent animals that run in the American West. Appealing stories and movies have been constructed about them. For a century ranchers have shot and poisoned coyotes to protect their animals, primarily sheep and particularly lambs. Today the coyote cannot be touched. Formerly coyotes subsisted on insects, rodents and small wild animals. Now, because sheep are entirely unprotected, they subsist mainly upon sheep when these are at hand. Lambs in particular are much easier to kill than insects, rodents and small wild animals. The coyote population likely will now expand rapidly and will consume increasing numbers of sheep, resulting in a probable increase in the price of lamb. Nevertheless, we should assume that coyotes will remain with their new dietary habits.

This transference of coyote dietary practices away from wild creatures and towards sheep is supposed to protect the environment. Such convoluted logic is beyond understanding. But it illustrates yet another point about regulation. Could it be that sometimes the justification for regulation is regulation itself? Especially when the ignorance factor is so high that rational solutions are beyond the perception of the regulator.

Two points could be strongly argued. First, rather than having to be forced into a certain pattern of health, safety and consumer relations behavior, it is in the farmer's interest to co-join all interested parties actively and positively to protect health, prevent accidents and to make consumers happy. The regulatory process, on the contrary, sets up an adversary process, a "we" and a "they" syndrome where parties involved can often become more concerned about the feud than the result.

Second, if positive action is indeed in the farmer's interest, why do things so often go wrong? Why do pesticides sometimes pollute rivers? Why do farmers fall off ladders once in a while? Why do consumers get meat with too much fat or a potato with a rotten inside?

There are two answers to this kind of question. One is that some degree of misadventure is part of the human condition. Regulators can regulate to their hearts' content, but can never stop a careless man from falling off a ladder. The other answer is that no one knows everything. There is no point in creating dissatisfaction among consumers by offering too-fat meat or rotten potatoes, no sense at all. In these cases one soon knows and the farmer adjusts, in his own interest.

Even assuming that mankind — in this case farmers — are totally venal and would use any information gained to play further tricks on the consumer, we question whether regulation would be the antidote. Assuming consumers are as gullible as farmers are venal, the only recourse in this case is to grasp the nettle. We need specific laws which, when broken, provide for prosecution and penalties to fit the crime. No more. If the situation is as bad as alleged, regulation is a mere cosmetic and a waste of everybody's time and effort.

This issue of regulation or service relates to the American experience. Circumstances differ from country to country, but basic considerations remain surprisingly similar. The differences tend to be of degree. It is almost as though bureaucracy were an international movement. In New Guinea, which in terms of culture and material development is far from U.S. standards, as one can get, one finds bureaucratic mores compatible with mores in Washington, D.C. It is as if there is a mind-set common to humans where, by similar method, it is "natural" for one human to regulate another.

The Third World in its need to develop its agriculture, has little if any need for regulation. The role of the U.S. Department of Agriculture is worthy of note here. Here is a department which, except for a few politically inspired *interregna,* has essentially offered service to farmers rather than regulation. Moreover, in the development of U.S. agriculture it offered precisely that kind of service now needed by the Third World. The Third World needs the opposite of regulation. It needs service from its sponsors, an educating process. The matter was put quite succinctly recently:

> **Does the proposed regulatory 'approach' [i.e., in the U.S. government] provide the least imposition on human freedom? Is it more productive to place the emphasis on opportunity and education instead of regulation? I can think of no better example than the unbelievable productivity of American agriculture, the envy of the world. It results to a major degree from the research and education done by the land grant agricultural colleges and the educational field work done by county agents. Where do you suppose agriculture would be today if the county agents had been regulators instead of teachers?[7]**

Here is the key and the antidote to regulations — to be teachers instead of regulators. This calls for massive decentralization so that teaching takes place on the farm. There is so much to teach. Let me offer a short example embracing six currently-known agricultural techniques, an example that could be multiplied many times. This "model" is designed to exploit human capital, brains and muscle, rather than money.

Soil erosion, pests and plant diseases are grave impediments to agricultural yields everywhere in the Third World. Simple experiments comparing mixed cropping to single crop operations have shown that the right mix of plants for a given location can curb and often eliminate erosion. It can greatly reduce disease and pests by reducing the natural habitat of these depredators in the rightly mixed crop situation. The mix will have to be selected according to ecological dictates of a particular region. No investment of money is required; just brains, imagination and communication.

A second technique suitable for peasant farming is overlapping cropping or interplanting. Plants grow slowly during early stages of their lives. Interplanting takes advantage of this by sowing seeds of a new crop between already growing plants, thus overlapping slow growth of one crop with faster growth of another. A typical program could be planting rice, then interplanting sweet potato, followed by an interplanting of soya beans, next corn, and again soya beans. As a result, one acre could yield, in one year, two tons or more of rice, ten tons of sweet potato, four tons of soya beans, two and one-half tons of corn, and another four tons of soya beans. The process is highly labor intensive and fertilizer intensive. Most importantly, it might also be called managerially intensive. In net gain one acre cultivated in this manner could most adequately sustain thirty people continuously. The capital need is slight, but the human investment is high, very high.

The third element in this simple "model" concerns the inescapable issue of fertilizer use. Agricultural development cannot proceed without simultaneous buildup of soil fertility and constant replenishment of nutrients which are used up. The use of chemical fertilizer is enescapable; such use makes agricultural and economic sense. But the amount of chemical fertilizer can be substantially reduced by proper crop practices. Such plants as lima beans or other native legumes can capture nitrogen from the air. If land can be fallowed and all or part of one bean crop allowed to rot in the soil, there is an appreciable nitrogen enrichment. This not only reduces other fertilizer use, but makes what is used more effective. Such savings and new productivities can be induced through education rather than regulation.

A fourth element applies to water-borne crops such as rice, the world's second largest grain crop. For many centuries wet rice farmers in North Vietnam have planted *azolla,* or water fern, in the *padi* along with the rice crop. As the *azolla* produces no seeds, starter colonies of the plant must be maintained year round, but this is not a major problem. The plant has many enemies and its care must be quite explicit. The water fern, *azolla,* has the capacity to fix carbon dioxide into sugar by photosynthesis. Along with this process the plant also fixes nitrogen from the air into a form that can be utilized by the adjacent rice plant. So farmers mixing *azolla* with rice crops can increase 100% to 200% over those of their non-*azolla* using neighbors. None of this creates need for giant manufacturing plants or World Bank loans. Rather, we see a potential to enhance the world's second-ranked grain crop through nature. These potentials are rare. When we find one, we should use it vigorously. Indeed, the cost return ratio is such that every wet rice *padi* should contain the remarkable *azolla*.

A fifth area offering similar benefit is the treatment of acid soils. Soil acidity is a world-wide problem. The normal method of treatment is to apply processed lime stone. Unless acid soils are thus treated, crop yields will be small. If animals are grazed on lime-deficient pastures, so too will their yields diminish. Research in Africa and England indicates that if leaves from selected trees are mulched and mixed with acidic soils, acidity is reduced so liming may become unnecessary. Again, the call is not for regulation, but educational service.

The last factor in this simple "model" is to make use of the more than abundant labor that exists in the Third World. There are a thousand examples. Herbicides, for instance, are a better way to control weeds than is hand hoeing. Hand hoeing loosens top soil and can lead to wind and water erosion. On the other hand, spraying weeds removes their demand for nutrients, but leaves the root structure to hold the soil. Hand spraying of individual weeds can reduce herbicide use, and therefore costs, by significant margins as opposed to universal spraying of fields by airplane. Where hands are plentiful, often a second plant, as given in our first example, can be planted in between rows of the main crop, not only hindering disease and insect infestation, but significantly adding to yields. Better use of labor, which is readily available, offers endless possibilities for increased production.

The real issue regarding this "model," however, is to put all these elements together in an integrated system. Each system will vary from place to place, but each system — except for additional fertilizers — calls for little more than brains and imagination. Beyond this simple "model" comprised of six elements, we can construct an endless succession of additional models, almost *ad infinitum*. This is not to suggest, however, that the Third World should be denied the "best." Sophisticated equipment, fertilizers, pesticides and techniques are imperative. What is suggested is that these elements be put to better use through application of human talent given in service to others. Instead, we create giant bureaucracies, allegedly to give service, but which instead, from their centralized remoteness, regulate with unremitting remorselessness.

We have already noted in another chapter a "Report to the Congress of the United States."[8] Regarding research and extension services, the Report notes:

Developing countries could greatly help realize their potential for food production increases by promoting their extension service programs and by devoting more resources to research on adapting new varieties and techniques to individual country conditions and needs.[9]

Although the same charge could be made in all 46 countries surveyed, albeit to a somewhat lesser degree, India exemplifies the failure of bureaucracy to stimulate human talent. By early 1972, Indian agricultural colleges had produced some 70,000 agricultural technicians. At that time, it was estimated that about 22,000 of these were totally unemployed. Another 30,000 to 35,000 were employed, but in New Delhi offices or, to a lesser degree, in state capitals. Probably fewer than 10,000 agriculturalists were in the field. Yet Dr. M.S. Suaminathan, probably India's leading agronomist, has stated that India's 100 million to 108 million ton grain crop could be increased to 210 million tons within fifteen years. This would require a national investment in energy sources for agriculture and in fertilizer plants. It is openly admitted that technical assistance, including loans, would be needed from outside.

The key factor however would be agricultural extension workers in the field. Yet in 1971, during the Bangladesh War, the transfer of American farm technology and technicians to India was broken off by India. In 1966, the Agency for International Development had 236 professional technicians in India, most of whom were agricultural specialists. The AID budget was $877 million in that year and most of these funds went to agriculture. A decade later the AID program in India had a staff of nine and no money. The Rockefeller Foundation, which had done yeoman work in agriculture on the sub-continent, had left altogether. The Ford Foundation which has a similar record in India, was a mere skeleton.

Yet agricultural expertise, especially in such esoteric areas as plant genetics and plant nutrition, is essential to India's education of agricultural workers. In areas mentioned, American expertise is supreme. Obviously the Indian farmer in the field needs expert help from his own kind. "Miracle" rice and improved wheat seed strains introduced in 1967 gave India near-sufficiency in food production by 1971. At the same time, it brought problems. We might digress briefly and discuss this issue because it illustrates both the kind of task sponsorship agencies face and it emphasises the need for service to predominate over regulation.

There is no doubt that "miracle" seeds produce greater yields per unit of area. Depending upon what base one adopts, these increases can be as high as 250%. But the new seeds produce new demands. Traditional rice varieties in particular have adapted well to widely varying water levels; they can survive in high water levels which "drown" the "miracle" sprout. On the other end of the spectrum, the "miracle" species cannot withstand even short periods without water. Their fertilizer demand is so constant, that without constant water supply to carry fertilizer almost continuously to the root structure, yields fall off dramatically. This calls for a modern hydraulic system very different from monsoon-dominated irrigation. To control water levels, there is a need for efficient pumps and drain-off systems. The need is vital. It simply cannot be neglected if the potential is to be realized.

Furthermore, the larger amounts of fertilizer required by the new strains of wheat and rice also create chemical imbalances in the soil, a new problem to the traditional farmer, although not to his Western counterpart. Excessive use of nitrogen, which new plants need in quantity, can leave soil deficient in potash. This is not a new problem, and the remedy is obvious, but one must have access to potash, and money to buy it.

New seed strains are less resistant to disease than old traditional varieties which have adapted to the rigors of life over centuries. Thus as regards "miracle rice" a demand arose for pesticides and that demand had to be met if the new plants' potential was to be realized. This called for access to capital.

As might be expected, the new strains required more labor. In dealing with nature one never gets something for nothing. An increase in yield calls for an increase in input somewhere else. Where a small farmer was already at his full labor potential (not a particularly prevalent situation), he had to hire labor outside of his family. A broad estimate of additional labor needed was from 25 to 50%. This is a significant increase, and to meet it could put an enormous burden on a small operator.

Therefore, the large farmer, especially those who had access to capital, benefitted most from "miracle" seed strains. Adequately financed, he could realize fullest potential of the new rice strains. Especially in India, it was the small farmer who needed to benefit. What he needed most was technical advice and assistance in the field. Peasants are enormously resilient and vigorous people. If they know how, they bounce back from adversity in extraordinary ways. Their capacity to improvise and meet a known adversity is often prodigious. However, they must know the dimensions of what faces them. This is the basic beginning, the first element of assistance in agricultural development. Such assistance had to come from a sponsoring agency. Knowledge was not available to the farmer through his own channels.

It should also be noted that new seed strains were more or less being forced upon the farmer. He was being told that with the same effort as previously exerted he was going to obtain a miraculous result. He probably did not believe this, having heard so many similar statements in the past, but the temptation, if not the direction, was there. It is also a pity that the appellation "miracle" was attached to the new strains. There are no miracles ahead of mankind in dealing with nature. But the new seed strains were a great simplifier. They offered to governments a seemingly easy solution to the food crisis.

Why did the new strains not live up to their potential, especially in India? It was not primarily a failure of agricultural research, and certainly not of farmer efforts. It was the failure of sponsorship. As the preemptive sponsoring agency, the bureaucracy had little desire and less capacity to give the service. It did not decrease its regulation. In India, even the irrigation system so critical to the proper use of the new strains is under bureaucratic control.

> **Timings of canal opening and closure are decided by the state, and the farmer has little opportunity to control irrigation depth, or dry the field in order to apply fertilizer. With the present water supply system individual farmers are often powerless to take discrete actions to improve their output.[10]**

To any farmer involved in irrigation the above situation is preposterous. In a particular field, it is vital that the individual farmer be able to control water according to his experience and judgment. This is fundamental. His water program may have to adhere to community needs, but without correct judgment in the art of irrigation, water and all kinds of effort are simply being wasted. At the same time as this situation existed, India had large schemes for extending irrigation. These projects, measured in acres to be serviced, have been impressive. In the 1960's more than two million acres per year were being added. But this seemed to be an exercise of form over substance.

That India's bureaucracy failed to sponsor agricultural development is not unique. In the GAO report, similar failures can be noted in the forty-six countries surveyed. Rather than rendering service, the bureaucracy is regulator, if not director. Speaking of Pakistan it was observed:

> **The government's use of extension workers as enforcement agents damaged their credibility with farmers. The agricultural attache said that the government had used extension workers to direct farmers to plant certain crops rather than only to assist the farmer. In addition, the GAO representative said that the conflicting duties undermine the potential effectiveness of the extension worker.[11]**

At the same time as this sponsorship breakdown is taking place, the state is not exercising other prerogatives that it has taken unto itself. The application of one ton of nutrient fertilizer to Indian fields is estimated to yield an extra grain production of five to seven tons. This ratio is impressive even by American experience. Indian fields are impoverished, thus results from any nutrient enrichment must be high. Nevertheless, a World Bank team, seeking to increase Indian fertilizer production, found fertilizer plants there were operating at a mere 60% capacity.[12] A comparable situation seemed to exist with agricultural machinery production. In reporting to his Board on operations in India, the Massey-Ferguson vice president for administration stated:

> **'Where we find the most frustration is India, ... we have a minority interest and no real say, and the government is notorious for its arbitrary actions.'**

Because of administrative bottlenecks, Massey operates at less than half capacity in that hungry nation.[13]

These snippets are but the outward indicators of the failure of bureaucratic sponsorship of regulation over service. There are even more serious problems inherent in the regulation process — extortion, bribery, the black market and smuggling. It is impossible to measure the scale of activity. In one instance, black-marketing of goods from abroad exceeds the nation's legitimate exports and imports. There are no figures for bribery and extortion, for obvious reasons, but it is widespread.

Bureaucracy is the focal point of the process, followed by the political process. Bureaucracy is awarded first place because it is larger; so extortion and bribery are simply larger operations. Although the scale of bribes and especially of extortion is wider in bureaucracy, specific amounts tend to be smaller. With politicians this is reversed. Bureaucrats tend to specialize in extortion and politicians in bribes. No assessment of bureaucracy as related to Third World agriculture is accurate unless extortion, bribery, black market and smuggling are considered. There is an etiology inherent in bureaucracy which draws bureaucracy to these malfeasances. Let us trace this etiology.

How are bureaucracies born? Let us assume that bureaucracies are born of initial good intention. The progenitors are politicians or academics. These national programs invariably require a bureaucracy to support them. Intellectuals see the programs as devices to better the human condition. Increasingly, they see themselves as policy makers who help shape the world into pre-conceived patterns.

Politicians are more direct. They see a particular program as part of their political capital. It is a means to promote their political tenure and power. Let us assume that these are well-intentioned men, even if tempered by self-interest.

Next, the activists arrive. Inspired by the program concept, they often want to help people, but they are not dreamers or romanticists. They believe in practicalities. These activists can create organization. These are persons who can drive other people to action. They get things done. They demand results. They want to see staff organizations, budgets, headquarters, reports, research, development and practical field operations. These persons could succeed in private business as well as in government. In the end, however, these activists become victims, especially if their programs provide bureaucratic solutions to human dilemmas. Bureaucratic solutions to any problem are seldom successful. Thus the activist, who is also a realist, must contend with failures.

We must remember that agricultural development is more a human issue than a technical matter... a human issue which reaches some fundamental human concerns. One's work methods, role in society; attitude to money and rewards, family relationships to work, family members' relationships with each other; relationship to the world, vision of the future; children's education, and relation to rising affluence — these are some of the attitudinal changes.

Bureaucrats as managers of this sort of change are severely constrained before they start. Indeed, they probably are unaware of the human complexity with which they must become involved. On the technical side, the bureaucrat also faces constraints. Unless he is prepared to rely heavily upon the peasants (and most Third World bureaucrats simply do not know how to do this), he can never acquire the parochialism and intimacy essential to the proper use of land. If he is a city-based activist, it is even more difficult for him. He must not only meet the localism requirement, he must sustain his interest for long periods. Through continual learning, in field work and research, he must make constant critical and ongoing adjustments. The activist faces a comprehensive, perhaps impossible, challenge under most Third World circumstances. For this reason activists seldom last long. They either depart or they lapse into the next bureaucratic stage in the etiology — entrenchment.

When a bureaucratic system becomes partially or totally ineffective, it does not cease to function. The more ineffective the bureaucracy, the more it grows. This is in direct contrast to private organizations which, under similar conditions, go out of business. For example, of the twelve companies comprising the Dow Jones Industrial Average in 1896, the year it was established, only two still exist. The rest were either replaced by a more efficient organization or disappeared because they were not needed.

This is not the case with bureaucracy. As noted in the "Parkinson Law," bureaucratic ineffectiveness requires an increase in the size of the bureuacracy. Ineffectiveness leads to reorganization. This can be quite an activity, generating its own rationale, if not its own dynamic. Operational activity related to the mission can cease; yet a new organization, new staffing titles and departmental names, and many other new processes will come into being.

These are not the substantive changes. Such take on quite another form. As actual field operations decline — as service declines — regulation increases. The trend is to operate, not in the field through service, but from the headquarters by directive. Regulation and all associated paraphernalia of reporting to, recording and legalizing to seek conformity, if not obedience, all become institutionalized. The bureaucratic mission becomes little more than reporting, recording and legalisms — regulation.

Bureaucratic regulation is a sorry thing, especially when applied to agriculture. When a bureaucracy regulates say, public health, it has taken an open-ended commitment. There is really no end to what might be done to support public health, short of abolishing all disease and death. This means regulation, bureaucracy, can grow indefinitely. Beyond adding new regulations, old regulations can be amended. These demonstrate in practice, as seems to be invariably the case, that the initial regulator did not know everything when he made his initial assay.

Another factor adds to the inchoateness of the process. No one is aware of all the regulations, least of all the regulators themselves. As noted, in 1975 for example, over 8,000 new regulations were added to the U.S. regulatory lexicon. Assuming that each new regulation required half a day's study, simple arithmetic shows that no single person can keep up with the rate of output. Thus, a dairy cattle feed additive was approved by one agency, but milk drawn from the cattle was condemned under different regulations against the additive, promulgated by another agency. In this particular instance, milk was poured down drains for nearly a year until the conflict was resolved — in favor of the additive. We can repeat, if all regulations pertaining to agriculture and agricultural work were observed, the mighty agricultural productivity of the U.S. would come to a halt.

This is especially pertinent regarding the Environmental Protection Agency and pesticides. Rachel Carson's book *Silent Spring* strongly influenced the way Americans view chemical control agencies. Banning of chemical pesticides became a leading tenet of the environmental movement. Chemical pesticides are highly toxic substances and their absence from America's fields would be good *if* something else could take their place. Fortunately something could. The new substances are biological pesticides, mainly insect-attacking bacteria fungi, hormones and pheromones (an attractant). The great advantage of these agents is that they appeared to have no toxicity to other forms of life, other than the targeted insect. They left the "good" insects alone. The biological pesticides are "natural" pesticides. For example, most trees exude a "natural" pesticide. That is why many tree species have existed hundreds of millions of years.

During the 1950's and '60's, American scientists made many brilliant discoveries in the field of biological pesticides. American farmers looked forward to their use. Then, in 1972, the EPA began to enforce the Federal Environmental Pesticide Control Act, as hasty and ragged a piece of legislation as ever came out of Congress.

A mass of regulations descended on the companies developing biological control agents. Each agent had to be licensed separately, even when there was only slight molecular change between members of a genre. Soon costs mounted, so that up to $1 million might be expended, before even one agent reached the testing stage. The cost of development became totally prohibitive even for the largest of companies. Biological control agents were quickly dropped. Commercial development of biological control agents were killed by bureaucracy.

The tale does not end there. Insects remain in the fields. Even the EPA realizes insects have to be controlled. The agency licenses chemical pesticides. There was a great insect infestation in the U.S. Southwest in 1978 because of heavy February and March rains. The EPA, in response to the insect problem, released limited numbers of toxic chemical pesticides for use by farmers. Had the EPA never existed, in 1978, we would have had biological agents to meet the insect infestation.

The whole issue is typical of a bureaucracy. The result is always a stifling of human effort and creativity, which often fosters an ill greater than that the original problem.

There is a built-in defense mechanism against all this waste, namely "benign neglect." Most regulations can be legally voided or only periodically observed. This is happening today in the United States regarding chemical pesticides. In the U.S. it takes attorneys at great cost to counter regulations. Legal fees become a "cost of doing business" and the unfortunate consumer eventually pays the bill. In the U.S. and elsewhere, everyone is dependent day-to-day, on agriculture. For the consumer, there is no escape from the burden of real cost.

The tacit understanding of the informal regulatory mechanism does not always work. A single bureaucrat at an agency itself may decide the regulations which must be implemented. Such a posture is inherent in the bureaucratic system. It hovers over any bureaucracy like Holmes' "brooding omnipresence in the sky." In this case a major problem arises. It brings us face to face with one of the major ills of contemporary bureaucracy, especially in the Third World, where enough impedimenta exist without other burdens being added. Where bureaucratic regulation physically impedes a system from functioning, almost immediately one can expect an incidence of extortion, bribery, black market, or smuggling. Such are the only methods left for getting around the impediments. The process deserves some examination because where it exists, it leads productive enterprises into new realms. An understanding of what transpires is essential if one wishes to come to grips with the real dynamics of agriculture, especially in the Third World.

Extortion is when routine payments are made to officials so that routine operations can proceed. Westerners who normally become quite concerned and moralistic about extortion and corruption should set aside such judgments. Corruption is a natural result of ignorance of the state and its failure to regulate trade sensibly, in a limited way. It is only through corruption that what should otherwise be unobstructed trade becomes possible at all. The ingenuity and enterprise of those who corrupt officials are immense. The cost is disrespect and contempt for law and its makers. It is to lawmakers and law enforcers that extortion payments go. The scale is vast.

When an extortion system has become self-regulating, it works quite efficiently. The basic principle is not to kill the goose that lays the golden egg. Bureaucrats who operate the system often are not paid any salaries, but are compensated on the basis of "reasonable" extortions. Any such system can hardly be applauded. It depends upon abuse of power, where small abuses tend to become large. The central government also tends to become involved in the system. Inevitably the system burdens chiefly the poor, from whose income it exacts a horrifying amount, sometimes 60 to

70%. No wonder the poor not only fear, but hate the extorting individuals. This means that officials, bureaucrats, can hardly be trusted as self-proclaimed agents of agricultural development.

Bribery, the second feature of the official corruption system, involves officials receiving money for allowing some specific act, or removing some bureaucratic impediment. It moves from project to project. It is also vast in scale — no one knows just how vast. There are probably no projects valued at a million dollars or more which do not involve bribery. This applies to foreign aid distribution, which is initially government to government, but must ultimately involve fund assignment incountry. Because politicians are normally decision makers, they are most commonly involved with bribery. Because they control police, give licenses to operate, can influence courts and allow media to function, they are very rarely charged with any crime.

Nor is bribery always non-productive. Much bribery money is well invested in productive enterprises, some of which form the economic base of the nation.

The third element of bureaucratic corruption is the black market. Bureaucracy can declare goods in short supply and fix artificially high prices. A black market for these goods will promptly arise. When, as in India, fertilizer plants run at 60% capacity, restricted by official red tape and bureaucratic controls, a shortage develops. A black market in fertilizer becomes inevitable. In Peru, low potato prices set by the government (ostensibly to make them more available to consumers) become unprofitable to grow, and a shortage develops. Accordingly, the price rises and a black market runs the price so high most people cannot afford potatoes at all. In Burma (where the black market is larger than the regular economy) the black market pays the farmer twice as much for his rice as does the government which by law is the sole purchaser.

The Soviet Union also has an extensive network of black markets providing a full range of consumer goods and of spare parts for machines supposedly used only by the government.[14] In Communist China the black market is supported, gangster style, by "speculators' organizations" complete with guns and supplied by thieves and underground factories. The bureaucracy in Red China is not part of the black market system, but makes it possible by over-regulation.[15]

Generally, the black market is created by ideology, which artificially interferes with natural economic forces. It is also a result of rhetorical promises of politicians, who can deliver on their promises only by introducing price and other controls which create shortages of goods and then negate the promise. Bureaucrats also cause it, by using their privileged positions to become parts of the regular retail black market. By so doing, they are very severely injuring the government of which they are a part. The black market pays no taxes, and by subsuming a great part of economic activity, it weakens the rest of the economy upon which the government ought to be based.

Regarding agriculture, sometimes it is the black market that keeps it prospering, by offering farmers a reasonable financial return for their labor. But ensuing breakdowns in controls must lead to ever more controls in a vain effort to rectify the situation. This threatens coercion, pure and simple. It also creates an attitude of short-term profits instead of long term development.

The fourth element of corruption is smuggling — another consequence of bureaucratic regulations. The huge effort marked by ingenious individual improvisation that characterizes smuggling, could much better be channelled into orderly economic development.[16] It is harsh and irrational regulation of commerce by the state which forces smuggling upon those who might otherwise be lawful entrepreneurs. It has little effect upon agriculture or farmers, who cannot well have the mobility to take any part in it.

The real price of extortion, bribery, black market and smuggling is the diversion of enterprise, skill and energy away from creating wealth and in thwarting political and bureaucratic strictures. Yet in some cases, such as Burma, the system probably prevents total economic collapse.

Still another characteristic of bureaucracy in action needs to be explored. The question is at what stage does growth in bureaucratic control become totalitarian? A simple definition of totalitarianism is monopolistic control over people's productivity, trade and commerce — control without any rivals. Such control over productivity, trade and commerce, must also mean a high degree of control over people's life styles and the way they think. It is a short but necessary step from control of material considerations to that of stark propaganda, control of media, severe laws and severe penalties against deviants, even to arbitrary arrest — and then to those ultimate forms of punishment characteristic of totalitarianism.

Such controls also have a further attraction to some governments. Carried to extremes, they can reduce consumption of material goods, thus lessening demand for all those informal activities. This is the way it is supposed to function and there is an attraction to it for those politicians and bureaucrats who deplore the informal processes of extortion, bribery, black market and smuggling, in which they see a threat to their own power and control. People see in a widespread system of corruption an attractive alternative to ineffective restriction by the state.

Indeed, those who initially reacted against state restrictions by instigating corruption often come to support those restrictions encouraging corruption. Restriction provides them with opportunity. This they seize. Where results are lucrative, they obviously desire to perpetuate the system. There develops a kind of stability, perpetuated by those who benefit. Sometimes the opposite is true. Restriction can breed instability, as inflation and subsequent easing of restriction allowing prices to increase to true levels. Thus ordinary people who have geared their life ways to artificially low costs are hurt greatly. Groups take advantage of this instability.

Even if they refrain from overthrowing a government, the contrast reveals a weakened state apparatus. These are compelling reasons for pushing regulation to a form of total control. Perhaps this was the purpose of a regulatory bureaucracy as opposed to a service bureaucracy — control, power over people's lives.

One obvious potential remedy would be to decentralize bureaucracy — to engage in devolution and at the same time seek ways of mobilizing human enterprise, skill and vigor. Devolution means to reverse other features. Instead of lessening freedom for the individual, it means an increase in such freedom, thus decreasing state power. This calls for a change in temperament of the state leadership from proto-totalitarian to public service.

The same applies to lesser mortals in government. The bureaucrat must move from regulator to servant, perhaps teacher. Extortion must go. Bribe-taking must become a memory. The black market must become a legal market and the ordinary man must be able to apply his enterprise, skill and vigor as he sees fit and on his own behalf. Bureaucratic ignorance born of isolationism must also disappear. One result of decentralization would be to have bureaucrats live in the countryside learning to live and work among peasants.

Whether temperament of the average Third World bureaucrat can bend to these changes is doubtful. Until it does, little will happen as regards sponsorship of agricultural development through bureaucracy. Instead, what virtually amounts to two societies will remain, the society of the city and that of the countryside. Nor can any other sponsoring body arise while a typical bureaucracy exists.

Bureaucracy cannot expect to be an effective sponsoring body while it is the focal point of corruption. The peasant simply will not believe that yesterday's extortionist is today's mentor.

It is conceivable that private groups could be established, specifically designed and skilled for agricultural development. For those groups to operate effectively, to have the autonomy to deal freely to develop an appropriate agricultural system, there would have to be an environment of freedom alien to current conditions. The reality is that bureaucracy is a formidable barrier to agricultural development. It cannot do the job itself. While it remains in its present form and exudes its present temperaments, neither can anybody else.

FOOTNOTES

1 James Burnham, *The Managerial Revolution* (Westport, Conn., Greenwood Press, 1972).

2 Comment and quotation from Weber in *The International Encyclopedia of the Social Sciences,* Vol. II, p. 207.

3 *Ibid.*, p. 213.

4 Norman Macrae, "The Coming Entrepreneurial Revolution," *The Economist*, Dec. 25, 1976, pp. 41-42.

5 *Ibid.*

6 Joseph A. Schumpeter, *The Theory of Economic Development* (Harvard Univ. Press, 1947), p. 151.

7 *The Wall Street Journal*, June 14, 1976, from a speech by J. Stanford Smith, Chairman and Chief Executive Officer, International Paper Co., "A Perspective on Regulation."

8 Report to Congress by the Comptroller General of the U.S., *Disincentives to Agricultural Production in Developing Countries*, Nov. 1975.

9 *Ibid.*, p. 14.

10 *Ibid.*, p. 56.

11 *Ibid.*, p. 65.

12 *Ibid.*, p. 58.

13 *Business Week*, Feb. 2, 1976, p. 44.

14 A. Katsenelinboigen, "Colored Markets in the Soviet Union," *Soviet Studies* (Univ. of Glasgow, Jan. 1977, Vol. XXIX), pp. 62-85.

15 Miriam and Ivan D. London, "Prostitution in Canton," *China News Analysis* (Hong Kong, 1046, July 9, 1976), p. 1.

16 Toward the end of the 18th century an inquiry was conducted as to why agricultural productivity was declining in the east of England. It was found that farmers along the coastal region were using their horses, not for agriculture, but for transporting smuggled French goods inland. Possibly 40,000 horses were involved.

CULTURE: BOTH INHIBITOR AND PROGENITOR 10

When we view culture and agricultural development, we enter a poorly defined area. First comes the word culture itself, which has become an escape word. When we cannot comprehend some interesting aspect of human behavior, we say that such behavior occurs "because of their culture." The word becomes a kind of bottom drawer into which we assign all human attributes which either seem to have a rational meaning or for which we do not have names or cliches, because they are uncommon. Before analyzing culture in relation to agriculture, we shall have to define it.

Culture is a body of beliefs, methods and habits held in common by members of an identifiable group and considered intrinsic to that group. They are passed on by family and intergroup association. The holders do not like to give up or change beliefs, methods and habits.

When cultural mores are disrupted suddenly or violently, great personal and group trauma can result. Methods and habits are intrinsic to us and we cling to them. They largely determine the way we look at the world on both macrocosmic and microcosmic scales. We cling to this triad of beliefs, methods and habits because it gives us signposts pointing a way to go, a path to tread. The triad offers us a degree of certainty in an uncertain and often chaotic world, and helps determine how we carry on our daily lives. This is the definition of culture in this book.

Because we are talking of agricultural change, culture as defined becomes a dominant factor. Peasants, the key to agricultural salvation of the planet are, like the rest of us, the inheritors of a particular cultural heritage. Indeed many would say that peasants, as a group, are culture bound. They tend to remain with what they know and believe in, preferring not to embark upon unknown ventures. Yet if agricultural development is to occur there must be changes of habit, method and slowly but surely, in peasant beliefs.

Culture is often an inhibitor of development; on the other hand, it is sometimes the opposite. We find cultural characteristics among peasantry that underpin development, not only because of the nature of the cultural underpinning relative to agriculture, but also because those underpinnings aid and give stability to human values which help individuals to manage change with skill and maturity.

Change can be a difficult, if not dangerous, process bringing with it not only potentiality for great personal and group trauma, but also for extinction of a given culture. To talk of cultural extinction is to embark on delicate semantic grounds. A culture is measured in terms of its most pronounced features, such as religion, system of governance and such social macro-organisms as community organization. The nature of its art and literature may seem to disappear in overt form. Seemingly, the units of the culture, its people, seem to phase out of history. It is a tragedy when this occurs, when change is so forceful that evolution of the old into something less old, but still recognizable, is overturned. Herbert Spencer noted that "Few things can happen more disastrous than the decay and death of a regulative system [he meant a code of behavior, not a bureau] no longer fit, before another and fitter regulative system has grown up to replace it."

Most of Burke's great treatises sounded even more eloquent warnings. Toynbee's master work, *The Study of History,* is basically the story of the decline, fall and rise of cultures. Yet here we are, in agricultural development, essentially gazing at some two thirds of the world's peoples — its peasants — and calmly stating that their agriculture, one basic root of any given peasant culture, must be transformed. It is a bold assumption, pregnant with all manner of disruptive potential, often shattering, surprises.

We can quantify agricultural change — the water buffalo is replaced by an agricultural tractor. Water buffalo in terms of draught power are simply too wasteful of energy. An internal combustion engine can return a 20% work efficiency on energy consumed and an electric motor, 30% to 35%. In terms of work, a water buffalo returns only 5% of the energy it consumes. With the replacement of the water buffalo, an entire sequence of other quantitative measures ought to ensue, each producing its own demand for adjustment.

It often works. It works best when a little is offered at a time. Perhaps with the water buffalo, what is first needed is a better designed plow and pneumatic wheels for the buffalo cart. We have the enormous problem of facing a great need, increasing productivity, with many, many optional ways to achieve the goal. It is a philosophical question, this quantification of agricultural development, and the choice of options open to us.

In our own society we see the same issue of options. One part of the world of science, the quantifiers who are so important to our quest for productivity, often looks with scorn on the introspection of philosophy, which is primarily concerned with knowledge arising from the study of principles in human action or conduct, which lead in turn to a system of ethics. Shakespeare put it well when he had Friar Laurance tell Romeo in *Romeo and Juliet:*

**I'll give thee armour to keep off that word;
Adversity's sweet milk, philosophy, To
comfort thee, though thou art banished.**

Other quantifiers in the sciences have a different view. We know that there were several thousand years of agricultural experience for mankind to use before any scientist defined and analyzed agricultural knowledge. Sometimes the traditional explanation — intervention of the water goddess as the monsoon rains flooded the *padi,* was just as useful as a treatise on meteorology. What meteorologist will assume that there is nothing beyond the monsoon, beyond the differential heating and cooling rates of large bodies of land juxtaposed to large bodies of water, subsidised by geostrophic wind forces occasioned by rotation of the planet? What makes the planet spin? Why does the planet exist? As in the opening sentence of Heidegger's "Introduction to Metaphysics": "Why are there essents [things which are] rather than nothing?" We should approach this issue of culture transferrence with a degree of humbleness, if not some respect for the considered opinions of those whom we have decided must change their ways. Very often, we find that ancient prescriptions are very practical, besides offering a satisfying balm to the spirit.

In Northeastern Thailand the monsoon can be quite erratic. The crop cycle can become a tense and quite desperate venture where every step must be skillfully executed if the crop is to succeed. We have already noted the trauma of "floating rice" cultivation. A young Peace Corps worker noted, however, that peasants in his area sometimes neglected their fields at important times while performing little ceremonies to the water goddess. The young man sought help from the Abbot at the Buddhist temple, a learned man, deeply involved in the peasants' work. The youth suggested that the Abbot call a village meeting where he, the young Peace Corps volunteer, would explain the meteorological phenomenon of the monsoon. This was done and all the elements were outlined to an attentive peasant audience: differential rates of cooling in land and water masses, geostrophic wind forces and the like. The villagers seemed impressed. The young man did not stress his point, that the water goddess was irrelevant to the matter, because he did not want to offend. But surely it was obvious. The rationality of science had to win out. As the lecture ended the Abbot sprang up and adopted quite a stern visage. He told the audience how grateful they all should be to the young foreigner for giving them this new knowledge.

Then the Abbot made his peroration. The whole issue of the monsoon, he observed, is extraordinarily complex, far more complex than even he had imagined. He had noted, the Abbot continued, that many people had become quite slovenly in their obeisance to the water goddess. This had to stop. In a matter as complicated as this, one could not be too careful or too attentive. So, go to your fields and to your shrines and make obeisance properly. The goddess must not again be taken so lightly.

Without doubt, however, traditions are superceded. There seems to be a natural process albeit a normally slow one. The imperceptible analytic capacities of a culture add new leavens. What the donor of new things must avoid is insisting that his beliefs supersede other beliefs.

Despite our belief in scientific objectivity, many of our own perceptions are still imperfect. They will always be imperfect and we should remember this. Science is in constant flux which today reveals the imperfections of yesteryear that were taken as truth. Traditional prescription also changes. Some tradition may reflect centuries of knowledge. Thus besides seeming to slow or prevent change, traditional culture may also offer a perch from which one can take off on a dangerous flight, yet know that one has a perch to come back to. There is an ancient Chinese aphorism which puts it well — "to have roots yet soar like an eagle."

What we seek is another concept of "mix," that goes beyond our previous mixing of techniques. We must be flexible enough to mix the seemingly modern and sophisticated with the seemingly ancient and serendipitous. It is not a task for the impulsive and impatient, or for men of certainty who know what is good for other men, where end justifies means. This aspect of cultural change, the forcing house of premeditated revolution, is also part of the scene.

We live in a revolutionary world — revolutionary more in a rhetorical, than actual sense. Revolutionaries, however, will have little patience with the idea of change expressed in this chapter. The revolutionary need not be a doctrinaire ideologue, nor an overt, field revolutionary as those, to quote Orwell's terrible phrase, who "Think in slogans and talk in bullets." He is likely to be a Western specialist whose ideas are just as fixed as those of the ideologue as to what must be done, and the rightness of doing it as a totality, now. To such a man, cultural respect is simply irrelevant. To him ancient culture is a massive impediment (as indeed, much of it is). If development is to proceed, he says, such impediments must be swept away.

We have sufficient evidence that when one tries to obliterate cultural mores, the holders of those mores are placed in jeopardy. In the process of "deracination," we witness rapid decay leading to death of all the old values and thus to a totally unpredictable future. The society may revitalize itself and recoalesce around some new set of values or, as has happened so often, it may simply die in a slough of physical and moral turpitude. What is also possible is that new values are rejected and old values persist under a different name. This seems to be an inbuilt defense mechanism in all viable cultures which, even under difficulty, shows a notable capacity to survive.

We have talked in general terms about the recipient culture with only passing reference to a donor culture. What of potential donors? These folk who are primarily Westerners belonging to a technological society, are seemingly the holders of great knowledge which must now be transferred. Let us let agriculture be the framework within which we make this examination.

As Western man has gained a technological temperament he has lost much of his agricultural temperament. In highly industrialized countries, except for Japan, agriculture is practiced by only a tiny percentage of the population. In the U.S. less than 5% are farmers. Agriculture practiced in industrial countries is quite different from what a small farmer in the Third World practices. What is practiced in the developed world bears only slight resemblance to what the Third World farmer might aspire to in practice. This gap is further compounded by the intensity of specialization in highly developed agriculture.

Today we find specialists who are enormously competent in their speciality, but who know very little about the totality of agriculture. Therefore to assume that a farmer from the West can communicate admirably with a farmer in the Third World simply because they both exact a living from the land, is far from correct. Both may have a "country man's eye" but they practice a different art.

Even among many farmers in the West, there has been a loss of temperament of the agricultural symbiosis because of agricultural specialization. To some degree, the peasant still deals with the symbiosis as a whole, even though he may not understand it. On his tiny holding he has to deal with plants in as many species as he can handle. He will keep as many animals as he can, and their nutrition, sicknesses, parturition and training are handled by him rather than by a coterie of specialists. He deals with the soil himself and is unlikely to be able to call on anyone to analyze or specify as to its better use. His lack of infrastructure, sophisticated machinery and irrigation equipment makes him even more conscious of the vagaries of climate.

When it comes to work, the difference here is probably greatest. It is astonishing what the peasant farmer has to do — and can do — with his hands, his eyes and his supple back. The products of a highly technicized and specialized industrial complex are simply not there for his use. If a peasant wants something he is likely to make it himself. The variety of farm tasks also falls upon him. It was not alway so — this gap.

A person who has lived the agricultural symbiosis in the West has to view the agricultural symbiosis of Africa, Asia or South America differently from one who has not lived it. But it is the opposite which is important. How does one who has not lived it see it — this urbanized person isolated from even the most elementary aspects of growing food and fiber? The best that such a person can do is to express an intellectual concern — reality has to elude him. Thus when urbanized Western man comes face to face with this problem of agriculture, he does so in an intellectual way. Then he turns to what he knows if he wishes to help with a solution. And he does not know enough. So he offers money to the world's peasants, which becomes known as aid. But it does not aid. There are no discernible differences after the money has been given. He turns to a "people to people" approach, but this is no more than a feeling gesture. His children are offered to the Peace Corps. How a young American adolescent barely able to make his own bed could positively effect the agricultural symbiosis among peasantry is truly a mystery.

The Peace Corps did, however, benefit young Americans. It provided a kind of crude psychic therapy for many. By becoming involved in the "you" of the world's poor they could either escape from, or, better, come to terms with, the "I" of oneself. But this is turning a peasant into an object to provide therapy for America's young. Surely that was never the intention.

There are yet others who offer different forms of self-abnegation. They would share with the needy what we affluent ones have. The need is too big to be solved by this. But it will effectively and rapidly result in a widespread distribution of poverty. Most others in the affluent West probably neither understand nor care.

In this overview of culture as it relates to the agricultural symbiosis, we can isolate two elements. There is a "they" and a "we," a Kiplingesque East and West which do not meet. "They" are the world's peasants, about three billion of them primarily conditioned by agriculturally-bound cultures. The "we," the urbanized minority of the world's peoples, know nothing of agriculture.

Yet there must be a relationship if peasant agriculture is to give adequate food, shelter and clothing to all and to provide, while doing so, a basis for further material and mental development. The urbanized West has a scientific basis. Science is indispensable to agricultural development. The West has developed management techniques; an adaptation of these is also required. The West has some capital liquidity and this is important. Particularly, the West's technology must be adapted and applied.

The West has something else more important than all these significant material factors. At the risk of oversimplifying, we call it a tradition of freedom. The tradition has burned low many times, but this cannot disguise the fact that material achievements of the West have been most successful where freedom has been greatest and individuals can rise to their potentials. Dictatorships have played (and are playing) a sorry role in the West in capital accumulation, scientific discovery, management and technological innovation and in production modes generally. What has to be achieved must be done by individuals. The freer those individuals, the more can be (and has been) done. The West has much to give in this process of developing human capital. Barriers to transmission are enormous, the cultural gap between "we" and "they" being the most significant. To a substantial degree, the West in particular has lost its rural roots and in so doing has lost a great deal. But the West's help is needed. It is a wise husbandman that knows the nature of the components he deals with before he begins preparing, sowing, growing and bringing to fruition.

The key to change in agricultural development is to work with culture rather than against it. Agriculture, especially in a peasant society, is something more than a job, a task covering eight hours a day where change might be easier. Agriculture is a way of life. When one changes agricultural practices, life styles themselves change. Those who would foster change must couple acute cultural understanding with caution that the prescriptions they offer must always take account of cultural reality.

One could, of course, ride rough-shod over cultural reality. But one had better be prepared to use force, and to the maximum. There is no need to dwell once again upon the mind-staggering horror of twentieth century efforts in this regard. Instead, one might observe what resulted. Did terror bring an agricultural cornucopia? Hardly, despite the agony. Andre Amalrik, the Soviet dissident, writes of his exile in 1965 to a remote Siberian village. He talks of village attitudes towards work, if indeed they could be called attitudes. Except for attention to their private plots, villagers avoided work. Amalrik wrote of their drunkenness, which was as constant as the supply of liquor would allow. In summary he says, "There is also the mentality of the kolkoznik (collective farmer): he has ceased to be a peasant but has not become a laborer, and hence cares nothing about what happens to the results of his work."[1]

Yet it was Russian peasants who, under the New Economic Policy of the 1920's, returned Russia to its prewar production capacity virtually overnight. If a government insists upon changing a culture rapidly, it must resort to violence. But after violence has reaped its grizzly ends, the situation is likely to be worse than before.

We have noted the persistence of traditional culture in the People's Republic of China. What we have witnessed under Chinese communist agriculture is a failure of collectivised agriculture and an uneasy return to ancient Chinese mores. Chinese communist agriculture today is different from the ancient *Ching T'ien* (well fields) and *Chun T'ien* (equal fields) systems only in terms of names and addition of better seeds, fertilizer and some machinery. With introduction of tiny private plots for Chinese peasants working on "collectives," the same phenomena as in the Soviet Union is developing: intense labor in the private sector, high productivity, and an immediate and positive response from city buyers in the private markets. Was all the violence, struggle and turmoil worth it?

Nevertheless, there is continuity of a certain cultural trait — the historically persistent drive of peasants to obtain the maximum share of their productivity. One witnesses it all over the world, from culture to culture. Its outgrowth in People's Republic of China (PRC) is particularly interesting in that there, new men were allegedly being molded in whom the collective interest was to supersede individual interest. It seems that the old cultural pattern really persists, despite the assault against it. Let us look at a refinement of this theme.

The long association between pigs and man is a root factor in Chinese agriculture. Mao Tse-tung understood this. Hog raising in contemporary China is outlined in a detailed, graphic article.[2] Hog numbers fluctuated wildly in PRC following assumption of power by the Communists. When private plots were taken away under collectivization in 1958, hog numbers "declined calamitously." Later, when private plots were restored, pig numbers increased dramatically. Theorists talked of the "absolute preference for common rearing" of these animals, but this proved impossible to implement because when the peasants heard these theoretical mutterings, the pig population dropped again.

Then came the cultural revolution, Red guards and turmoil. This had another devastating effect on the pig population. Peasants sold their pigs. There was an over-supply, the price of pork fell, and peasants slowed the production of pigs as typical supply-and-demand reaction. In 1970 newspapers began to quote Chairman Mao on pigs. "One man, one pig, one *mow* [a unit of land], one pig." The Fukien Daily stated: "every pig is a small organic feritilizer factory." "More pigs more fertilizer, more fertilizer, more grain." "More pigs the more meat — thus it will be possible to save food grain."

This sensible admonition had to be implemented. Here contemporary Marxian theory and ancient Chinese life styles came into conflict. The radicals had opted for collectivization of pigs, but "the mere mention of collectivization has slowed down the growth of the pig population." There seemed to be a good reason for this slowdown. The state as buyer simply did not give peasants an economic price, so they responded as peasants have always responded. They cut down the supply of pigs, which was relatively simple because of the animal's short gestation period. If it does not pay, we will not do it. The ancient style persisted.

By February, 1976 the pig situation had reached another stage. The Hunan Daily opined that the policy about pigs should not be slowed, because the policy had been established by Chairman Mao. "Collective rearing of pigs should be promoted, but the present policy allows individual rearing, which, therefore should not be restricted." The distinction here is not overly subtle. What is being said is that the Chinese peasant, with overlord and bureaucrat, had once more won out over radical theoreticians.

No doubt the drama of pigs vs. collectivization has not ended. The stake is the most important animal in China, the pig. Once again we witness the persistence of cultural habit. This has always had a practical side in agriculture.

But there were difficulties in the other China though perhaps not of the same magnitude. On Taiwan the strength of Chinese culture is also demonstrated. That culture is so old — and within it the peasant has endured for so long — that it simply has not gone away on the mainland because of theories developed by a 19th Century German intellectual. Neither did it disappear on Taiwan in the face of more rational attitudes by authority under less stringent circumstances.

The scene is a meeting of a Farmer's Association in the Republic of China in Taiwan.[4] The meeting had been called to settle a particular matter. The vote was 51% for and 49% against. Disorder arose. There were two points of contention. Why, it was asked, should a decision be made this way? It was unfair to the 49% who thought otherwise. Such a small majority should not rule over such a large minority. The voting division arose, in fact, out of ancient clan divisions dating back many generations to life on the mainland, each clan accusing the other of wishing to impose its will.

Still other ancient patterns emerged. The peasants turned to the local magistrate who was present. You adjudicate, they insisted. You make a decision and we shall abide by it. It was the way it was done in the past. The magistrate declined. It was the government's directive, he said, that these matters were to be settled by majority rule. The farmers could not accept this so they turned next to the Senior JCRR official present. You decide, they said. He declined for the same reasons as had the magistrate. They then turned to an official from the Provincial Department of Agriculture. The result was the same. Someone then found what he thought would be the solution — let a visiting American decide. The American, Dr. David Rowe, explained, since he was only there as an observer, it really was not his business. A wise decision, one would think.

The meeting broke into disorder. Majority rule is perhaps acceptable when a large majority overwhelms a small minority, for the minority will normally come over and join the majority, producing a consensus, another ancient Chinese posture. When the split is more or less even, majority rule has traditionally been viewed as unfair to one side. This would certainly be a Chinese view anywhere the culture persists. The solution is for some person in authority to adjudicate — a father relative to sons, older persons relative to younger ones, or an official relative to citizens. That has always been the relation of government to people. Yet Republic of China policy has been to withdraw as arbiter in agricultural operations, and to give farmers those powers. It would seem that culture sometimes gets in the way of good intention.

Moving forward in this synopsis on culture and agriculture, we must observe also another feature. There are differences in culture. Everyone knows and accepts this truism. The important point is that cultural difference is reflected in agricultural difference.

The Malaysian war lasted from 1948 to 1960. The insurgency was led by the Malayan Communist Party, which was ethnically Chinese. The war was fought in jungle and mountain areas. The Chinese grew crops in their remote base camps to help sustain themselves. In the same remote areas were many indigenous people who also grew crops for themselves.

From aerial photographs, it was simple to identify every Chinese garden from the way it was laid out. The Chinese system of ridging and trenching was quite unmistakable. The value of this to the British in their suppression campaign is obvious.

Perhaps, however, the most distinctive difference between cultures in action as expressed in agriculture comes from Micronesia. A grouping of islands, Micronesia has more than 2,000, dotted over 3 million square miles of Pacific Ocean. They total approximately 110,000 square miles of land. The 1970 population exceeded 100,000 with seven main ethnic groups. During WWII the main islands, trolled by Japan, provided some air and naval bases. The U.S. liberated the islands in 1944 and 1945. The group is now divided into the U.S. Territory of Guam, the government of the Northern Marianas, and trust territories under U.N. Mandate with the U.S. as trustee. The exception is the Gilbert Islands, which the British control.

Under the Japanese, the whole territory literally groaned under its agricultural accumulations. It exported food to Japan, even though the wartime population was greater than today's. Now the area is food deficient and imports food — even canned fish. The soil has not changed. Climate is the same. Plant life is as it has been for centuries and the ocean still abounds with fish. War's impact in any physical sense has long been healed. Its only legacy is that Saipan is attacting Japanese tourists who want to visit the scene of one of the war's most bloody battles and the place where Japanese women and children committed suicide by the score rather than fall into American hands.

What has changed is agriculture, then and now, has been in the hands of two different cultures. During Japanese occupation, Japanese assisted by Okinawans farmed the land. The natives were but onlookers. Now responsibility belongs to them; or is it to us, the American trustees? Under Japanese occupation Saipan had a flourishing sugar cane industry which exported processed sugar to Japan. Today there is no sugar cane, no sugar industry. Under the Japanese, allied with sugar production, there was hog farming. The hog population was about 350,000. Today it is about 12,000. Similar pattern can be found on all the islands, embracing all the ethnic groups. The disparities apply to other foodstuffs as well as hogs and sugar cane.

The situation has not changed, nor is it likely to change, under the American aegis. In a report to the Secretary of the Interior, an American agricultural expert has observed:

> **The ratio of food imports throughout the territories, compared to exports is better than five to one. In some areas such as Yap it is closer than ten to one. Yet throughout the districts that I visited I found a universal shortage of many foods, along with seeing vegetables, fish and other foods rotting. With this high ratio of imports and exports, the cost of living as related to food is extremely high to anybody who does not have a subsistence farm. Unfortunately the food surpluses seen rotting were caused by encouragement to farmers to grow more without solving the problems of marketing and transportaion. This, in my estimation is the quickest way to keep these people farming at only subsistence levels.[5]**

Micronesia is a classic case of culture triumphing over technicality. There is no technical reason why Micronesia lacks a flourishing agriculture or even aqua-culture. But the indigenous population traditionally saw no need for this. There were fish in the sea and when fish were needed, enough could be caught. One farmed on a subsistence basis, supplemented by wild growing plants. To many Westerners with romantic notions it was an idyllic existence. It allowed for the cultivation of other values besides toil on land and sea, and Micronesians had their own intrinsic values in plenty.

Today, under American stimulus, the situation is changing. Little Micronesian talent is being directed towards agriculture, fishing or aquaculture because politics and bureaucracy are attracting the bright young people who normally would provide entrepreneurship. Under American stimulus, consumption is rising rapidly and products are more varied and of higher quality. Micronesia is today further away from being self-sufficient than at any time in its history. It promises to be a ward of the U.S. indefinitely. This is immensely important to many Micronesian leaders, and of equal concern to U.S. officialdom.

Micronesia is a microcosm of a much larger world problem. It offers another example that cultural values tend to persist, rejecting change. Micronesia also illustrates that a culture will flow away from the difficult and toward the easy. Under American patronage Micronesia has become a microcosm of the ultimate bureaucratic state, where talent moves toward government and away from productivity. As long as there is an external source of wealth willing to finance such illogic, the schemata will continue. Micronesia is tiny in terms of numbers of people and the U.S. is large and rich, so in this limited case a *modus operandi* can be established. But this does not presuppose that this is a *modus operandi* for the rest of the world or indeed for any major national grouping. This draining of talent, away from productivity into negative productivity, living on what another produces, is a trend throughout the Third World.

History shows that agriculture was adopted slowly and reluctantly. Even at the beginning of time, people preferred to do something other than painstaking, difficult tasks of husbandry. Today there is an attitude of rejection — a belittling of agriculture as beneath the educated and ambitious. And the system continues as long as the ignored peasantry remain with their fields and animals, producing sufficient food and fiber, albeit at minimum rates. What will the ultimate end of these cultural attitudes be? What kind of cultural respect remains to a people who have avoided optimizing their own production potentials? This question faces not only the Third World. It is equally real for the developed world, where many forms of productivity are ignored or misunderstood. At root it is a simple question of widespread economic illiteracy that seems to grow unaffected by contemporary educational modes.

Up to this point, in generalizing, it has become clear that culture is a pervasive force in any situation where change is a feature. It must be dealt with as a factor whether one sets out to promote change "by insensible degrees," or to accentuate it by coercion or force. Agricultural development occurs only when that development adjusts and relates to cultural reality. This is not to say that culture only inhibits change. On the contrary, if given opportunity, it may promote change. Let us look at other cultural factors in greater detail.

This book has emphasized the agricultural symbiosis. The first element in the symbiosis was plants. Let us look briefly at plants and culture.

Rice is the second largest grain crop in the world. Some 362 million tons were grown in 1977/78, compared to 382 million tons of wheat. More people are directly engaged in growing rice for their own subsistence than there are growing wheat. Rice is usually consumed as a whole grain, whereas wheat often is ground into flour. The taste of rice, then, becomes highly important in a culture.

We should never underestimate taste. The simple taste of what we put in our mouths is an intrinsic factor in how we judge things. There is also a high degree of habit in taste. We become accustomed to the taste of things. We are highly suspicious of, and reject, things that taste different. Since we eat two or three times a day, taste is constantly with us. We are more or less indoctrinated by what we taste all of the time. Not only indoctrinated in terms of our taste buds, but also by external factors, such as "What is it? Where did it come from? Why does it look *that* way? Is it clean?" The first time an American is served dog meat in a Maiou village in Southeast Asia illustrates the point. Dog meat is treated like any other edible meat in S.E. Asia. To see that small tongue (the best part, the Maiou insist) and a piece of rib, and to know in came from an animal that Americans *never* eat, but which is a friend and confidant, makes eating dog flesh very difficult.

But to return to rice and taste — various kinds of rice differ widely in taste. This becomes very apparent when rice is the main part of one's diet for any length of time. It can vary from large barley-like grains grown under "dry" upland conditions, to soft, sticky, glutinous, quick-growing "floating" rice. Taste varies likewise in the types between. The rice eater becomes quite accustomed to the particular taste of his rice.

When "miracle" rice was introduced, it very often faced a taste barrier. When a peasant rice grower cultivates rice, he does so to provide first to satisfy his own needs, which means satisfying his own taste buds.

This is quite different from farmers who grow market crops and buy food from a store. There are tomato farmers who dislike tomatoes and dairy farmers who do not drink milk. Such a rejection by tomato and dairy farmers' taste buds does not curb their husbandry! They continue producing and selling. With the subsistence farmer, the situation is exactly opposite. The plants he grows are directly related to his taste buds, for he grows primarily for his own subsistence.

There were loud complaints about "miracle" rice's different taste. The Westerner will exclaim, "How can a desperately poor farmer reject the benefits of increased yields because of a fine distinction in taste?" The Westerner forgets that *is* the distinction. Few subsistence farmers can yet visualize selling their entire crop and purchasing other foods which please their palates. They do know that to be safe, they have to eat what they grow, and taste tells them what is good and what is less good.

The Westerner should not judge this attitude too severely, for he, too, is vulnerable. One reason Westerners often find some difficulty in socializing with peasants on a one-to-one basis is because of *their* taste buds. If one is in a situation that calls for intercommunication between a Westerner and a farmer of some different culture, then a capacity to eat that culture's food becomes critical. If one cannot meet in the headman's house and enjoy a simple meal with him because the food is so strange, "repulsive" or "dirty," one cannot communicate. Particularly, if one rejects food because one suspects it is "unhygienic" an immediate barrier to normal, civilized intercourse arises.

Further, if one cannot eat the food, one cannot travel far. In peasant societies there are no clean, formica-topped, chrome-bound, stainless-steel-utensiled cafes. If one wishes to be mobile, one must eat the local diet under local conditions. Yet Westerners often find this hard to do for much the same reasons as so many rice farmers prefer to eat the old variety as opposed to the more productive new. But taste buds adjust, albeit slowly.

The second element in plants and culture is the way in which persons conceptualize about plants. In the American tradition, grass is a romantic concept. The "sea of grass" that comprised the Great Plains was intrinsic to the expansion westward. Around it was built the romantic legend of noble Indians; simplicity of right triumphing over evil in cowboy tales; and the notion of an untrammeled opportunity for the individual to live autonomously. The "sea of grass" is a phrase that any herdsman would understand, even if he were the first domesticator of cattle in 10,000 B.C., probably taming the giant wild ox, the Bos primogenitas. It would make little difference if he were an ancient hunter following the game herds. Or if he were that pioneer foraging westward across the Great Plains, with his horse for power and cattle for milk and protein, all dependent upon grass.

There would be little difference if he were today's Wisconsin dairyman viewing his Holstein herd in late spring. All could understand the emotion and perceptions stimulated by a Sea of Grass. A Javanese householder can be dismayed by the behavior of his bamboos, that marvelous plant which produces food for man and beast and all manner of uses for its stem. The latter provides building frames, house floors, fences, utensils, furniture, cages, toys, whistles and other musical instruments. These are but some of the uses for bamboo. Once in every human generation, from 33 to 36 years, the Java bamboo commits suicide. A bamboo which has reached considerable size, will suddenly grow enormous flowers, using up the plant's food resources until it dies. That which was so useful becomes useful no longer. The Javanese peasant is impressed by this phenomenon and some ritual observances surround the event. It demonstrates both the willfulness and the independence of nature in the face of man's needs and desires.

There is another kind of relationship to plants in human culture. Modern, urbanized man seems totally oblivious of his daily dependence upon plant life. Plants are objects of little significance other than perhaps beautification. The herdsman with his satisfaction at his sea of grass or the Javanese in awe of the explosion and death of giant bamboo (and in a hundred other illustrations) express a sense of primal relationship with natural phenomena. It is a relationship in which man is the object of the phenomena and not the subject. Western man has lost this sense of awe and he will never regain it. He needs a new relationship. The proper conceptualization of plants in a technological culture should be that of science and economics. This is not to negate the drama of the "sea of grass," the symbolism of bamboo suicide or the sense of sheer beauty and elegance that plants can give to man in endless variety. These remain and always will, as long as man thinks and has emotions. To project plants on the basis of human life, however, we need to go much further. Here technological man can render a great service. His science and economics are the two critical elements in this concept.

There are some ten million plant species on this planet and each one is a tiny, intricate, precision chemical and solar energy factory. Without exploring botany, it is still possible to emphasize that a scientific conceptualization of the ten million species leads to a capacity for potential development important to man's well being. Such scientific research is full of exciting possibilities for human betterment. It has become culturally unfashionable to laud science in the West. It is nothing more than that, however, merely unfashionable. The reality remains.

It is only through scientific conceptualization leading to scientific and economic action that the mechanics of agricultural progress become available. In the event, the science involved may be quite empirical as it was during most of the great agricultural advances in 18th Century England. The scientific approach, empirical or pure research, represents something more important than methodology. It represents a state of mind, a cultural attitude, in this instance toward the mutability of a plant and optimization of its intrinsic character. It is the acquisition of this state of mind which presents the mystery to us. How does it come about: this seeing things differently, the adoption of a scientific viewpoint? This will have to be explored.

A similar situation relates to economics. The basic economic viewpoint is that of striving to optimize plant productivity, to get the maximum marketable yield for least energy input. It is a state of mind, a cultural attitude. Both situations, looking at agriculture scientifically and economically, engender something else. It leads to a dynamic state in which the agricultural process is not static but changing.

To be sure, much of the old remains, for things connected with plants, animals and soils change quite slowly. Again it is a state of mind that is dynamic, not static. In terms of the economic state of mind there is a relationship here between culture, politics and bureaucracy. Will the political and bureaucratic system allow the farmer to obtain his share of the benefits of economic optimization? Without adequate reward the opposite will take place. It has always happened that way and in the most recent instance, a report from Vietnam illustrates the point. Commenting on the proposed collectivization of agriculture in the Mekong Delta it was said,

> **A major stumbling block is peasant conservatism in the south, lack of incentives in the North...individualistic peasants in the Delta [i.e. of the Mekong River] however are often reluctant to grow more rice than necessary for their own requirements and tax. They see no benefit in producing large surpluses that must be sold to the state at a low price or when there are few consumer goods available to be bought with the proceeds.**
>
> **Out of fear that excess family holdings of rice paddy would be collectivized, many farmers in the Mekong Delta divided their holdings among relatives and transformed paddies into ponds or orchards. Blaming such action on a misunderstanding of official policy a Saigon newspaper *Giai Phong* said this would 'cause considerable difficulties in mechanization of agriculture.'[6]**

Thus the development of economic attitude, an attitude toward optimization is often clouded by external factors, sometimes totally destroyed. The same could be said of personal attitudes toward science. Yet in the development of human capital, new, optimistic cultural attitudes by individuals are vital needs. Each farmer needs to grasp them and make them a part of his cultural matrix.

Neither is the distinction between science and economics imprecise. On the contrary, we are seeking two quite different and new cultural attitudes toward two related things. Such new scientific and economic attitudes form a basic starting point for those who believe that the planet can support itself admirably in terms of food and fiber. It becomes an article of faith that the scientific and economic mind set must predominate, without having science and economics shut out all else. This needs to be part of the cultural matrix of the West as it approaches agricultural development.

When we move our cultural exploration from plants to animals we find yet more cases of how culture can guide, shape and often inhibit attitudes as to the ways in which things are done. No better illustration of this can be had than by viewing the place of the cow in culture. Cows have played a substantial role in human history and still do so. They are remarkable animals. If ever we populate another planet a cow should be one of the first creatures to make that journey. There are approximately 135 million beef cattle in the U.S. and 12 million dairy cows. Beef and dairy products provide more than half the daily food for Americans, and most nutritionists agree that they provide balanced and essential human nutrition. Humankind can live without animal products and many humans do. In fact vegetarianism has become quite popular with some Westerners. Nevertheless, the cow is with us as bovines have been with us for some ten to twelve thousand years.

The cow has great symbolic value. Gandhi said "The cow is a poem of pity. She is the mother to millions of Indian mankind." A Hindu text admonishes, "All that kill...cows shall rot in hell for as many years as there are hairs on the [slain] cow." The cow thus became the symbol of nature itself. What a curious symbol the cow is. Its odd shape makes one laugh. The fore legs do not seem to belong properly to the body, but to be attached as an afterthought. The head attached to that scrawny long neck, is much too large, as are the ears. Then there are the horns, bony protrusions in variety of shape. The body itself is all angles, bumps and bulges. When the animal runs at a gallop it looks ridiculous. Only the eyes are redeeming. The cow reminds one of the whimsy of existence.

Beyond whimsy, however, the cow has been accorded something else — a sanctity. It reminds humans of their dependence on nature. A human could live entirely on milk. Sustenance like this comes from nature, for which man should be thankful and humble. His frailty in nature is obvious and mother cow is there with her funny shape to remind him of his frailty. A cow in some minds therefore becomes a sacrosanct animal. This concept goes beyond India's veneration of the animal. The cow as Nature is found in other cultures.

There is an Icelandic legend from ancient Norse mythology which also gave to the cow a significant role in human creation. Sutur was the first spirit. "When the lost smoke from Muspellheim met the ice clouds of Niflheim, water drops formed over the empty gap between them. By the power of Sutur the drop quickened into life in shape of a human, a being called Ymir." The legend then continues:

> **Out of the same clash of fire and ice Sutur caused a cow to be formed, Audhumla, the Nourisher, on whose milk Ymir lived. Audhumla herself existed by licking the stones [legend does not specify by whom they were created], which were covered by salt and frost. As she licked there sprang forth the hair of a man, then a head, and on the third day an entire man beautiful and strong.[7]**

The man, rejoicing in the name of Bur the Producer went on to great deeds including the siring of Odin, one of several sons who became the first Gods and Creator of the World.

In Medieval Europe the cow was an important if not vital animal with honor among men. The castrated male, an ox, was primarily a draft animal. From the cow came a rich variety of milk-based products such as cheese and its by-product, an amber liquid called whey. Milk was drunk fresh. It was soured and made into many different foodstuffs. Yoghurt, one of these derivatives is still popular.

In U.S.S.R., Stalin's collectivization plans almost foundered on the issue of the cow.

> **The first years of 'collectivization' were full of massive and persistent riots of the womenfolk which were caused by what Stalin later called, our 'minor misunderstanding' with the Kolkoz women, 'about the cow,' certainly an understatement trying to play down what really amounted to a major social contest between the peasantry and the State.[8]**

In any event, Stalin lost. In 1935 the measure which legalized private plots in Soviet agriculture also included a private ownership of a few cattle. In the following years these animals provided the bulk of Soviet meat and milk.

Bovines are intertwined with human history. Primarily they have given man bodily sustenance besides pulling his plow and his wagon. But the cow and the attitude of humans toward it reveal something more. Around the cow have sprung up cultural attitudes which limit its productivity. Some of these are deeply rooted.

The greater bulk of that enormous, vibrant society that we call India, pregnant with possibilities, venerates cows. India has between 200 and 300 million of them. These animals are not eaten. They die of disease or old age. Some pull carts and farm implements. They produce a small, insignificant amount of milk. In the full flush of milk production, shortly after calving, India's cattle might produce daily about 10 to 20 pounds of milk. In the U.S. specially-bred and properly-nutritioned milk cows can produce one hundred pounds of milk per day for extended periods. In India, even the animal's dung is not returned to the soil. Usually it is gathered dry and used for fuel. There was once (and still may be) a subbranch of the untouchable caste which followed cattle to pick through the dung for undigested seeds. These seeds were washed and eaten.

Yet, these Indian cattle, the Brahmin species, are magnificent animals, especially if one puts them in another cultural context. Genetically the strain is extremely stable; other breeds have had little interrelationship with the Brahmin. The breed itself has existed for more millennia than anyone knows. The Brahmin has been the primary animal in establishing a tropical and subtropical beef cattle industry, especially in the lower latitudes of the U.S. The animal is "heat resistant" to a degree unusual in bovines, and more disease-resistant in tropical and subtropical climates than European breeds. The animal is an excellent forager and has less fat (less waste) than other cattle. It is slower maturing, some advantage under certain intensive nutritional conditions. Nevertheless, some 200 to 300 million Brahmin cattle add nothing to the Indian human food chain. They are, agriculturaly speaking, a negative factor. They deplete grasses and other plants, with no return. A sizable human force must care for them when they are old. At times they block traffic. We must now offer a contrast with countries that do consume beef and milk.

We have noted the disparity between milk production in India and America. The ration noted was approximately ten to one, but it is actually greater. United States' intensive dairy cattle feeding keeps that animal milking longer and more productively than under "natural" conditions. Indian Brahmin cattle have not been bred for milk. Indeed, they have not been tampered with genetically at all. Appropriate line breeding, allied with better nutrition, could significantly increase Indian milk production. Improvement in nutrition would require a reduction in the country's bovine population. This would certainly mean bovine slaughter, either shortly after birth or in some other way as retention for beef production. Culturally this is unacceptable in India.

In terms of beef production, the contrast is more graphic. The object of raising beef cattle is to grow animals to their optimum weight as quickly as possible, as economically as possible, with minimum waste or fat.

In the United States a 1000 pound, eighteen-month-old steer will produce approximately 450 pounds of edible meat. The hide will produce about sixteen pairs of shoes and other assorted leather goods. The hair, horns and hooves, being protein concentrate, are processed and fed to hogs and chickens. Hair from the ears is used for brushes and tail hair is used for several special purposes.

A significant portion of the animal's internal organs, including intestines is eaten by humans. The remainder of intestines becomes pet food. Up to 400 other ingredients are recovered, most used for medicinal purposes. Dung may be used for fertilizer, although increasingly roughage (which is essential to a bovine's digestive system) is being recovered from the dung and fed back to the animal as part of its daily required roughage intake. Annually Americans consume nearly 100 pounds of beef per person, attesting to the usefullness and efficiency of the beef industry. It is also worth noting that beef consumption escalates with affluence. As incomes rise, so does demand for protein — beef protein, if it can be had. Milk, other dairy products and beef are intrinsic to the American diet, especially for the young.

The contrast between the positive impact of cows upon the American food chain and the situation in India is striking. It is the contrast between the cow as a productive animal on the one hand and as an animal to be venerated on the other. It is the contrast between two cultural attitudes. But it should not give rise to invidious comparisons. If Indians in their culture wish to venerate cows, that is their choice. Westerners, after all, give near anthropomorphic qualities to cats and dogs and would be horrified at the thought of eating these creatures. That is the Western cultural attitude. Nevertheless, expressed in simple agricultural terms, can any nation, India or any other, afford to support 200 to 300 million non-productive bovines which consume so much? The agricultural answer must be "no." In cultural terms, however, the answer would be "yes!" The cow has a place in Indian culture. If not, the cow would have disappeared from India long ago or would be raised for human consumption.

There are still other cultural attitudes towards cows. In great segments of Africa cattle raising is the primary industry. These various African tribesmen love their animals. The language used in addressing them tells us that. In some regions cows are eaten. In others they are kept for their milk and their blood, both of which are drawn frequently, mixed together and consumed. Nutritionally it is a good diet. In parts of Africa a bovine may be valued not for weight or milk, but for size, shape and symmetry of its horns. Compared to the West where the animal is evaluated as a production unit in the food chain, the contrast becomes striking.

Similarly, in Africa the value of a herd of cattle is not in the beef or milk production, but in the number of head owned. With such values, herdsman will graze as many animals as possible on a piece of land. This leads to overgrazing and other ills. A slight drought or too much rain can produce a disaster when land is over-grazed. Even under the "best" conditions overgrazing results in low milk or meat yields. This is one reason many bovines in Africa produce no more than 50 to 80 pounds of meat after five to seven years of growth.

There are other, more subtle, cultural attitudes towards animals. Once this writer was asked by an Asian cabinet minister how to start a cattle industry in his country. The minister was an extremely urbane man, with a Western education and long association with the capitals of Europe and the U.S. He wanted protein for the populace. There was available land and money. The minister suggested that the finest of pedigreed cattle could be imported to start things off.

It was suggested to the minister, however, that there already exists in his country an admirable basis for a beef industry. The country is a natural habitat for millions of water buffalo, which are the primary draft power in the nation's agriculture. It was explained that at any given age they attain about 50% greater weight than any Western breed, and would yield about 50% more meat. They are excellent foragers adapted to the nation's grasses. No imported animal could do so well under similar conditions. Further, in tropical conditions, water buffalo are more disease resistant than a Western breed. There is also very little fat on a buffalo which allows higher meat yields than a Western breed.

Provided buffalo are killed young — perhaps at eighteen months — not after ten or twelve years of work, the meat is tender and flavorful. Surprisingly, the water buffalo, if properly husbanded, is an excellent milker. It can yield 40 pounds of milk daily, with double the butterfat of a Western breed milk cow. Most important, it was explained to the minister, buffalo are not only native, but farmers are familiar with them. Each animal has a status comparable to that of a family member.

The biggest advantage would be to establish an organization, probably a cooperative, to handle large numbers of animals as opposed to the peasant's one, two or three animal herds. The minister listened to this program patiently. He similed and then equally as patiently added that I had not understood. He was trying to bring his country into the modern world. What was being suggested was not modernization, but looking backward. The minister felt, "only poor people eat water buffalo!"

American stud cattle were imported, all fine animals. Artificial insemination, difficult to handle under tropical conditions was attempted in spite of warnings. Money was spent on experts and facilities, but the project languished. The peasants, who had been promised genetic miracles through inseminating their native scrub cattle from these massive Western bulls, were disappointed. Nevertheless, it was true that their imports were the finest cattle from the West, a symbol of modernization. Whereas those buffalo — well, every peasant has one of those and only the poor, who were obviously backward, would look upon them as a protein source.

Again one faces the same issue as arises with cultural attitudes toward plants, "How does one conceptualize a situation differently?" But we must note a refinement. The refinement of the question is not to say "How do we get them to conceptualize as *I* of the West conceptualize?" India is a large, exciting land with great human talent and an enormous potential. It is wrong, however, to expect most Hindus to conceptualize a use for their cattle comparable to the role assigned to Western cattle. This Western assertion leads only to cultural confrontations which have little to do with the question of how to grow more food. The Hindu cultural attitude toward the cow is too deeply rooted to be toppled simply by proof of the cost effectiveness of cow slaughter. India has a major problem with its cows, but the solution has to be an Indian solution bearing a strong relationship to the way an Indian views the world and his place in it. Perhaps to venerate the cow as a symbol could be achieved with far fewer animals but certainly not through eating them.

Nevertheless, cultures *do* change, albeit by a mysterious process. If one relates cultural change to his own lifespan it is obvious that revolutionary change accomplished in one's lifetime has a justification which may not apply to evolutionary change over centuries. It is a short step from this to declare that the end justifies the means and that any revolutionary excess which moves "change" forward is not only warranted but is one's duty to implement. After all, man has but one life: he should use it effectively by attaining a goal. People may be resentful and resist, but these folk represent the debris of history. They can be swept aside for the good of the many, and especially for the good of posterity. Indeed posterity and its welfare can justify anything.

Despite the seeming logic of the revolutionary, history shows that he does not move culture "forward" any faster than culture wants to go. Does this mean then that humanity is caught? Does it mean that in this instance the Indian culture of the cow must prevail because culture-change marches slowly and perhaps to no particular end? I believe not. We all need symbols but symbols mean little as symbols if they in turn overwhelm us. The cow can be a symbol, if that is what a culture wants, but only a few are needed for this. India in its future may never eat beef. But neither need India's 1977 632 million population compete for livelihood with 181 million cattle. Nor need the cow disappear as a symbol.

Perhaps, to be practical, more cows should be spayed and bulls castrated until the symbol reaches a symbolic size. The Indian case is perhaps the most complex and important example of culture's intrusion into agriculture. All cultures have their own peculiarities; India is not being singled out as an exception. Each one of us is a child of his culture and it colors our viewpoint. Indians are far from unique in their view of life, bovines and society.

We know that in addition to reacting to animals, culture also reacts to soil. Previously we observed the near mystical quality given to land possession. The same theme in Tolstoy's short story could have arisen from many different cultures. There may be a territorial imperative regarding land. It is not claimed in this book that this imperative is a dominant in man's social conduct, let alone an imperative in history.

Nevertheless when one remembers the minutiae, arguments and passions that arise from boundary disputes, one cannot ignore man's sense of territorial integrity. Incidence of passion is no less with nations than with individuals. Probably more disputes occur between individuals and groups along boundaries than in any other single locale. Where there is close land settlement cultures develop complex social structures supported by rules, to prevent disastrous disputes.

Probably the best example is the island of Bali, Indonesia. Many steep hillsides in Bali are terraced to support wet rice cultivation. Each cultivatable terrace area is small, sometimes only 40 or 50 square yards. Most of these terraces are individually farmed, thus cooperation regarding water control is essential. When terraces are flooded, capricious individual behavior, especially on the higher reaches, has devastating effects. Balinese terraced agriculture offers an example of a culture providing constraints and opportunities to allow a complex agricultural system to work. Elaborate ritual, if not semi-religious observances, comprise the cultural fabric whereby the complex Balinese terrace irrigation functions.

The Balinese system illustrates the human capacity to handle what appears to be an impossible situation regarding intensive land use by many people. Territory must not always be fought over. A culture can handle that. It is not an imperative that some foreign theorist's ideological model gives the framework which must shape men. On the contrary, the Balinese illustrate that complex joint land use can be handled by humble people. Perhaps that is the key. The land-use system in Bali evolved without ideologue, politician or bureaucrat. It grew culturally and has a definition, a sanctity that preserves it as a working system.

On the Aran isles, off the west coast of Ireland there is a different pattern regarding man and soil. The isles are little more than huge rocks jutting from the Atlantic Ocean. Yet, relative to their ability to provide sustenance they are quite densely inhabited and have been more so in the past. The great stone forts of prehistoric inhabitants still remain as do the ruins of a once-elegant abbey, for some three centuries the home of Irish monks. The ordinary inhabitants, until the tourist phenomenon burst upon them, were fishermen who also farmed tiny plots. There is little soil, sometimes none at all. Yet potatoes, which became a staple probably about the 16th Century, require a relatively deep tilth. This was created by human labor where soil did not exist before. Kelp was gathered from the tiny beaches at the foot of great cliffs, and along with some sand, was carried to a site by baskets. The site had to be hand-cleared of rock which was broken by hammer and mixed with kelp and sand. The kelp rotted and finally a soil emanated. Potatoes, with their extremely high caloric value, offered a very effective utilization of this area.

The Aran experience illustrates human necessity for soil. Without it, the Aran Islanders had only two options: to eke out a precarious existence from the sea and depend on an exclusive seafood diet, or to leave. The soil created mainly by human hand offered a fuller, different life. Animals were added once soil came, once plants developed. Perhaps it is unnecessary to worship the soil. Reverence, however, is not out of place. How many cultivatable square yards of soil could a person create in a lifetime?

In America there is yet another phenomenon regarding soil. Land is plentiful, and still cheap compared to most other parts of the world. Here is a situation which will never occur again. This phenomenon gave rise to two seemingly conflicting trends. First was a tendency toward rapacious, wasteful and disrespectful use of land. This trend continues, perhaps not so much in agriculture as with urban development. In the U.S. it is a cultural trait to build on flat land, in sequences of spreading rectangles, and to attract urban people by offering them an acre or more as their home site. This acre, however, does not grow the family food. The phenomenon might be called the "Los Angeles syndrome."

Then there is the "Manhattan Island syndrome" where people are piled on top of one another as were ancient Hohokam cliff dwellers in the Southwest. It is assumed that it is cheaper to populate flat areas, but depending on the time frame, this can be disputed. In the longer term it may prove to have been cheaper to populate the hills around the Los Angeles basin and to have left the flat lands as productive farms. The hillsides might also have offered advantages. An intense population density would have offered one way, through proper soil containment, of preventing the often disastrous erosion which now plagues many southern California hillsides. More importantly, the variegated land forms of the hillside would have broken the rectangular pattern and given a dimension of difference, if not intimacy, to living patterns. But the rectangles march on, changed only by circles and ellipses which do give some variation; but essentially they follow a pattern of urban sprawl, devouring prime agricultural lands. The same pattern exists in countries as diverse as England, Nigeria, Brazil or Australia, to mention a few.

The plenitude of U.S. land has also created a new reverse situation. The vast unfarmed areas of the 19th Century stimulated a farming revolution, combining applied technology and science in farming practices and general agricultural education. As has been noted, this essentially American technique has limited application elsewhere, although single elements built into new and different systems have immense significance for world agriculture.

This American venture has produced several extraordinary results. First, scientific advances have created enormous advances in unit productivity. This has arisen from plant and animal genetic research, soil research and equipment development. There have been similar advances in husbandry techniques, including management. This in turn resulted in the world's highest productivity per farm worker. Some 4 to 5% of the U.S. population are farmers who produce food for the world's most varied diet and produces 90% of the world's grain exports as well.

Culturally, this land is important because most Americans take land and productivity for granted. This has arisen not only from the efficient use of land, but from the plentitude of land and its cheapness. Most Americans simply are not involved with land use. Perhaps this is why the environmental movement is so strident, emotional and at times so enormously ignorant. Its members simply are not involved in any realistic sense. Like the others, they do not really know how things work. This allows passions to run riot.

Even this situation does not really alter the picture of a huge, highly efficient agriculture about which most of the populace knows nothing. Culturally, Americans are urban-oriented and have ambivalence toward soil. Impoverished peasants may have a deep attachment to the soil. Who could not when he has so little, yet is completely dependent upon it? Who could not when the pattern of his community life is dominated by his use of a little soil in harmony with the same use by many others? Who could not feel some attachment when he and his ancestors *made* the soil on a rock-bound island?

Yet these same folk have difficulty conceptualizing the role of soil and its future. They have virtually no scientific — or even general — informational input to their thought processes. Their conceptual horizon is further bound by limited availability of soil, science, technology and general information. Perhaps because soil appears to be inanimate, humans have greater difficulty in relating to it than they do to plants and animals. Certainly this appears to be especially true among urban folk. Land is something to build upon...to develop. Possibly the vast bulk of the world's populace needs a new cultural outlook regarding soil, its potentiality and its limitation to support human existence. The peasant farmer probably understands this better than his non-farming technological brother, but he is still detached from modern scientific reality, that supporter of soil.

Neither will education make much difference. When we are in doubt, we advocate more education, and just that. If one knows what to do, one does the job, rather than offering general theories as a substitute. Education in its general sense can help form a new attitude regarding soil, but first must come the state of mind; then education will follow. That is a cultural process which fosters and encourages especially the young to see a reality. Soil is the habitat of plants which in turn support the rest of human and animal life. Soil exists in limited parts of the planet and everywhere is a relatively thin skin. It was born of climate, wind weathering, water weathering, of differentials of heat and cold, to say nothing of an Ice Age or two. It was born of plant life itself to which soil gave rise in the first place. Soil has a complicated explanation in physics, in inorganic chemistry and organic chemistry. Soil is created every day and dies every day. It puts things into a geological time frame which men usually do not understand or appreciate.

At this stage of agricultural development it is probably better to look at soil in area terms also. It takes a certain amount of soil to support one human life. In history four to seven square miles supported one ancient hunter-gatherer. Today, one acre in Taiwan supports seven people. The situation is dynamic. The area of soil to support one human life is linked to the hand and mind of man. Here is the source of mysticism, born of the concept that this frail uncertain creature, man, should be able to achieve so much with soil. For we are now partners, soil and man. Whatever is to be done to support human life in the mass, soil and man must do it together. Not many cultures look upon soil this way. To do so opens new vistas upon which the continuance of life depends.

Climate also calls for different perceptions. The role of climate in the agricultural symbiosis is obvious, composed as it is of temperature and the availability of water. These two factors allied with soil properties are intrinsic to all plant life. But climate has an autonomy of its own. Man may improve the genetic structure of plants and animals, often with dramatic results. Soil can be nurtured by man's husbandry, even made by man's muscle! But climate is subject to only a little modification. Irrigation — capturing, holding and redistributing water, is perhaps man's greatest achievement in handling climate. The practice is very old.

> **It was...disciplined corporate effort, with a religious faith as its inspiration and with the necessary political authority and technological equipment at its command, that reclaimed the Afrasian river basins and valleys for agriculture. Unless a markedly different rainfall and weather pattern could be postulated for four or five thousand B.C., which we doubt, extensive life in alluvial Mesopotamia would have been literally impossible without irrigation.[9]**

New kinds of irrigation, which pump water from deep underground Ice Age aquifers is today only partially exploited. River diversion will increase. The planet has ample supplies of fresh water, but it is distributed in such a way as to make optimum use difficult. We must have water redistribution. Mechanical and other equipment exists to achieve this.

Man has attempted temperature adjustments in a rather inchoate way...not so much in terms of indoor and hydroponic farming as in capturing and moving water into areas conducive to good farming. The ideal system probably would be to irrigate the world's deserts. Here the temperature spread encourages high productivity of a large variety of crops, although by no means all. The humble potato, which along with rice, cassava and manioc, gives the highest calorie return per unit area and is the fourth largest crop in the world, does not prosper in hot, irrigated deserts. Nevertheless many crops do. Further, farming in the desert can proceed pretty well year around, resulting in higher yield per annum than in areas with short one-crop seasons. This blending of transferred water with favorable temperature is highly productive. But the technique, along with irrigation itself, makes only a small dent on the arbitrary rule of climate.

Man has long recognized the arbitrariness of climate. We have briefly mentioned the Water Goddess. A whole sequence of reverential customs could be cited. To make obeisance before a little-understood but obviously powerful force is neither uncommon nor unnatural. It may lead to humility which in the end is the basis of wisdom.

Even climate, or its personal impact, can be adjusted to, somewhat. Presumably the first human to don a warm garment did just that. This requires a certain state of mind, something between abject humility and an irrational search to dominate Nature. The key is finding the "mean."

Recently a church raised money to help drought-stricken Sahel villages in northwest Africa. The money was divided among a large number of villages, each receiving a small sum to improve each village well. It was a genuine "people to people" approach. But in terms of adjusting the water condition of the villages it was entirely wrong. There was no possibility that a $1,000 to $2,000 well improvement could ever alleviate the devastation of a long-term drought. Instead, such limited well improvements more likely tended to rivet present inefficient and insufficient water systems more firmly on each village.

Had the total sum been spent upon one deep-well irrigation system, a small area of land might have been adequately irrigated for several centuries. Under the Sahel there are immense aquifers. This small-scale example of a true climate modification was not what the church group wanted, partly because they sought an emotional contact with the largest number of people possible and partly because they really did not understand the dimensions of the issue. They could not establish the "mean" for this particular circumstance. Attaining that understanding is what culture change is all about. We are all bedeviled by our lack of it. It comes down to an ability to conceptualize based upon accurate information and knowledge. But each time is different. One can approach soil in a fairly pragmatic and broadly scientific way and anticipate a given result. With climate the approach must be different. Climate cannot be molded by man. The best man can do is adjust to it. This approach calls for a new cultural attitude. So we inexorably build a matrix of such attitudes to the agricultural symbiosis. It is from the influence of this matrix on our thoughts that action will produce effective or ineffective results.

As far as climate is concerned, we must return to a state of humility. In the end climate will win. A new Ice Age or, conversely, a warming trend will throw all agricultural projections into an entirely new context. And climate changes will come — geological history tells us that. The best we can offer is optimized skills, enterprise and courage.

The last element in the symbiosis is work. Of all the elements, the approach to work is most important. It is the dominant. Here we see very distinct cultural attitudes, for whether they seek it or not, men do not take work lightly. The first attitude is attitude itself. Why is some work more highly regarded than other work? In particular, why is field work regarded as something less than work on a factory assembly line?

No doubt many theories can be given to answer these questions. There seems little doubt that denigration of farm work rests not so much in the work itself, for the work is variegated; it calls for commendable skills, often allows one to enjoy extremely pleasant natural surroundings, offers the satisfaction of tangible results and is always interesting. But farm work has suffered from interaction with the city. Until advent of mass media, country folk were cut off from stimulus of city entertainment — I hesitate to call it culture, for that puts culture into a different sense than we are discussing here. The city gives a greater sense of opportunity and personal mobility than the bucolism of the farm. In the city one is at the center of action.

There are still very practical disadvantages for farm life in most parts of the world. Often transporation services are poor or non-existent. Health services and education for younger children are lower than in the city, even in most developed countries. While these disparities exist, practical reasons for preferring urban to country living will persist. Yet, it must be blindingly obvious that the great peasant mass cannot all move into the cities. There is simply nothing for them to do and nowhere for them to go. Also if they left the land, the cities would starve.

This is self-evident when one observes the great slum belts around Third World cities, inhabited almost entirely by deracinated peasants. With only a tiny fraction of the world's peasantry affected, we already have a disaster of mammoth proportions and immense tragedy. More importantly, peasants are needed on their farms. Their potential for effective work on those farms is enormous. Through productivity arising from effective work, they offer the very bedrock upon which the civilizing process might advance. Thus the first advocacy — or is it a plea — regarding cultural attitudes to work is directed less to farmers than to city dwellers. It is to accord due status to farming. The accord must be practical and attitudinal. Practical in terms of offering farmers and their families, wherever they are, normal services. Attitudinal, in giving them honor for what they do. This is a first, but basic, step. Let us return to more practical matters regarding work, for there are several distinct cultural traits essential to effective work in agriculture.

The first is linked to a word which is not sufficiently valued in the contemporary West: discipline. To be a productive farmer, to seek maximum plant and animal productivity, one must practice discipline. First, certain steps must be followed to grow things properly. These must be done on time and must be related to a particular requirement. There is no room for neglect, or slovenliness, or "doing one's thing." This applies to all effective work, farming or otherwise. With farming, other dimensions make discipline imperative. The farmer is involved with biological functions. A carpenter can take time off from a half-finished house and resume work later — or not at all. A plant or an animal is alive and its life is a biological function which adheres to a given time frame. It cannot be put aside or stored while animate — at least not for long.

A kind of finesse is required as one seeks to increase productivity through discipline. For example, cows will produce more if they are milked at the same time each day. Irrigated crops must be watered at precise intervals, not when someone feels like it. Soil that bears plants must be fertilized on schedule and the harvesting process must be geared to plant or animal life cycles.

These and many similar factors are basic to agriculture, and call for discipline. Greater finesse in production calls for even more discipline. This is an inexorable law ordained by nature and unlikely to be subject to change by either nature or man. Of more importance, the discipline must be self-imposed. If it is not, but is imposed by some external force, the whole system seems to break down. Self-imposed discipline has to have a cultural root. In China and Japan where work discipline is obvious, the family system also aids the process. Both societies are unilineal descent groups with clear, definable relationships. The family forms a hierarchy of values, roles and obligations and we must assume that its induction into society induces a disciplined attitude. One does not do as one feels but acts according to one's role.

In the West, especially in contemporary U.S., such attitudes are far from popular. Yet U.S. agriculture is productive and disciplined. Perhaps this is why only 5% of our people are so involved. But few other societies can yet contemplate such a situation — this assignment of a tiny minority to society's most important function. The point is that discipline, especially applied discipline essential to agricultural activity, has to be learned. It can be learned in the family, in the educational process and in the fields themselves. It is an inner state of each individual. Without it the individual cannot succeed in optimizing agricultural productivity. There are areas with undisciplined farmers, but these usually live off Nature's bounty, otherwise their days as farmers are numbered.

Another critical cultural and work factor in agriculture, and a major incentive to self-discipline, is money. There are very few non-money using communities in the world. As opposed to a century ago, virtually all peasant economies are now money economies. Money becomes both a measure of success and incentive to succeed. The single element to which peasants respond most readily and positively is money.

Pre-money economies undoubtedly responded to the idea of reward, and sought it in one form or another. There are few societies that do not recognize that work should be rewarded. Even in societies where leisure has a high value, reward for work is sought. Therefore, any culture which neglects or impairs this simple and obvious fact will have difficulty in optimizing its agriculture.

We have noted the dynamism of the small private plots in Soviet and communist Chinese agriculture. These plots did not match the ideological thrust; they were contrary to the official ideology. They were a bastardized expression of a universal cultural norm, that farmers should be rewarded for their effort. To refine the concept, if the farmer puts forward a greater effort there should be a greater reward.

In addition to the two main communist ideologies, there is a trend in the Third World to ignore, if not refute, the notion of increasing reward. The alternate notion goes under a vagary, that is, the same reward for all. Such equality is often cloaked in such phrases as "redistribution of wealth" and similar rhetoric. What does such redistribution mean if equality of reward is to be more than a mere slogan? It means the end of unequal reward, that is, reward according to individual effort. All cultures, if left alone, agree that effort should be rewarded. Money is now the measure of reward. Those who deny its attraction fly in the face of what seems to be a universal cultural norm.

Then there is the cultural attitude toward cooperation in work. In most cases this also embraces the role of the family. There are a few instances, and only a few, where a farmer can work alone. The contrary is the norm. Not only is cooperation needed intra-farm but in the best situations it is needed farm-to-farm. In the latter case a generalization can be made. For small farmers to obtain the capital, intensive equipment and goods needed as part of agricultural development, they must cooperatively purchase and share these items. This is the key factor in assessing cultural attitudes to cooperation.

Generally, cooperation within families can be depended upon to work.[10] The significance of family-type cooperation, however, is the carry-over of what is learned in the family, to the act of cooperation in a wider community. The inner cohesion of the Chinese family system can be carried over into wider associations. Chinese secret societies, which have a long history in China, are organized basically like a family structure. The same applied to craft and commercial guilds and to clan associations.

Today in the Republic of China on Taiwan the enormously successful Farmers Associations reflect the Chinese family system. Where they do not, and we noted one case, the Associations have trouble. When, after World War II, trade unions developed in Japan, the traditional obligatory system between younger and older persons in village society, the *oyabun-koyabun* relationship, was found to be part of the way the new Japanese trade unions functioned. While this traditional relationship might not have been what the American architects of Japanese trade unions intended, it proved to be a splendid lubricant to make the alien schemata work.[11]

In Africa we find a somewhat different, but also responsive, tradition regarding farmer cooperation. Many of the ancient tribal traditions required a high degree of consultation between all members. The purpose of tribal discourse was usually to find out how to act together. (This is a generalization because these cultural attitudes vary from group to group.) Overall in Africa one can sense a strong and positive capacity for cooperation among people. Such cultural attitudes have to be judged on their merits to each cultural grouping. Each potential will be different and will require different treatment. The organization of a Farmers Association in Taiwan, admirable as this may be in an operational sense, might not work if precisely copied in Africa, or vice versa. Cultures differ, and the inherent potential of each for cooperation must be evaluated according to individual merit.

The tragedy of attempts to set up cooperatives to date is that the fundamental issue of cultural differences has been ignored as though longstanding cultural traditions did not exist, or could be trampled upon. Organizations were imposed from outside or from on top. It was organization *per se* that was to stimulate productivity. In most cooperatives or communes the organization was supreme and the individual was subordinate.

Yet all experience indicates that the near reverse must be the case if cooperation is to be successful. The individual must reap an individual benefit through the working of the group, and the group must work according to their capacities and needs, the group benefits. But the imposed cooperatives, collectives and communes do not see it that way. Rather, organization becomes a device to control the lives of individuals. Suffice to say, capacity to cooperate comes from culture. It may be transmitted through family. It varies from culture to culture. Agricultural development, especially in the Third World, cannot occur without cooperation. The organization of such cooperation, however, must be based upon intrinsic culture values to begin, let alone grow and evolve. It can be taken as an axiom that the reason imposed cooperative systems do not work is that they seek control over individuals rather than enhancement of cultural tradition.

These words on work and culture are commonplace. It is quite old-fashioned to talk of self-discipline, individual reward, cooperation, family. The trend is toward self-actualization, which means doing what one likes. Individual reward is regarded as immoral and equal reward as virtue. The family is regarded as redundant. Elitist dominated organizations, "for the good of all" replace cooperation by individuals. In agriculture it is possible to measure the consequences of this current trend. The result is stagnant productivity and a slowly increasing impoverishment of billions of persons who in good faith listen to the nostrums of the educated and powerful.

How does one summarize relations between culture and farming? Culture plays a somewhat mysterious role in agricultural development. In one sense it binds people to the old, to that which must be changed. But if the old is snatched away, the ensuing trauma can be destructive. In other circumstances cultural traits blend perfectly into the new, and form a base for production advances. We simply do not know enough about the relationship of culture to development to make general pronouncements which hold good in all cases. We can recognize, however, that culture exists: a pervasive force which must be recognized in any development schemata. Neither should we lightly toss aside that which seems inhibiting, for there is much that is a progenesis. We must learn how to evaluate culture critically. We must then adjust and compromise, with gentleness and respect. We all need cultural symbols to learn of revelations of ourselves and of others. Symbols show us the joining of humanity in its seeking to be on terms with nature.

Many years ago my younger daughter and I arose early one New Year's day and went to Pramane Ground. Pramane Ground is a large tree-lined oblong of grass and is one of the focal points of the city of Bangkok. It was a Sunday morning. Already the grounds were encompassed by vendor stalls and marketing had begun.

We walked across the grounds to a small group of monks. These monks had many cages containing small birds. They were rice birds, normally considered a great delicacy in Chinese restaurants. We purchased a cage of birds. My daughter opened the cage door and let the birds fly away. Other people were doing the same thing. It was New Year's. In so doing one symbolized reverence for all life and of the need to come to terms with its mystery.

I thought, as I should have, of part of the Prakat Deveta, a New Year Supplication — "May men, or four-footed or two-footed creatures, not be subjected to danger this year; and may rain fall bounteously this season on all of the land."

I looked at the monk with his cages — and hoped that the Deveta might come true. The birds had long since disappeared toward the great rice plain.

Culture has its place. Culture must be accorded its dues. The task will always be to give it a role which facilitates, but does not inhibit. There are no rules. Beyond awareness born of experience, there comes common sense and respect.

FOOTNOTES

1 Robert G. Kaiser, *Russia: The People, the Power* (New York, Simon and Schuster Pocket Books, 1976), p. 96.

2 "Pigs, Resistance to Collectivization," *China News Analysis* 1037 (Hong Kong, April 16, 1976), pp. 1-7.

3 *Ibid.*, p. 3.

4 Recorded by Dr. David Nelson Rowe, a close friend and mentor, who attended the meeting and knows as much about the Republic of China's remarkable Joint Commission on Rural Reconstruction as any other living American.

5 Jack P. Rubel, A Report to the Secretary of the Interior: "Trust Territory Islands of the Pacific: Food, Fiber, Protein Action Proposal;" Phoenix, Arizona, May 1974.

6 *Far Eastern Economic Review,* "Hanoi Comes Down to Earth," Feb. 4, 1977, p. 30.

7 Katherine Scherman, *Daughter of Fire* (Boston, Little Brown & Co., 1976), p. 113.

8 Lewin, *op. cit.,* p. 2.

9 Arnold J. Toynbee, *A Study of History,* Vol. XII (London, Oxford University Press, 1961), p. 339.

10 Even here, however, there are exceptions. For one example and certainly an isolated one of non-family cooperation and indeed of no cooperation at all, see Colin Turnbull, *Mountain People* (New York, Simon & Schuster, 1972).

11 See James G. Abegglen, *The Japanese Factory* (Illinois, Free Press, 1958). However, Chiaki Nishiyama, *The Price of Prosperity* (Hobart Paper, 1974), suggests the carry-over is not all that significant.

11 TRANSFERRING WEALTH

The inhibitors of agriculture noted here have a much more pronounced effect on traditional agriculture than upon more developed farming. This chapter concerns development and underdevelopment in agriculture measured in quantitative terms, and how to transfer productive agriculture to the underdeveloped world.

This transfer relationship must be more than an agricultural relationship. A main point in this book is that technical improvement arises from human improvement, from development of human capital. Without such development there can be no optimum material development. Thus, while eschewing the pejorative insinuations inherent in dividing the world into two simple categories, developed and undeveloped, neither should we avoid the essential human focal point.

In general, developed nations can set the stage in several important ways. The first element is to know the limits of what can be accomplished by the developed. Much of this book is concerned with those limitations. Most of the effort must come from the underdeveloped themselves. But the developed can help make that effort more effective. The first element here is an exposition into truth.

To begin, we must assess explicitly agricultural potential in terms of prospective yields of a given country or region. This is rarely done. Instead aid often begins with vagaries. The donor wishes to "increase the production of rice" or "introduce new garden crops," or "promote hybrid corn." These statements do not specifically show to a country what it really might do.

Most countries would be surprised, even staggered, at the agricultural potential they possess if a true agricultural balance sheet were prepared to tell aid recipients in strict quantitative terms and on a national scale what is possible. Such a balance sheet ought to stimulate imagination and lift aspiration toward what specifically is possible. It is essential that this be a practical, non-ideological statement. Do not promise miracles, new eras, new worlds, or new men, or building this or that kind of new society. Let old societies remain. They will adjust in their own way, in their own time. Thus a country may remain extant rather than become a fabric torn by old passions and new prejudices.

There are many aids to assessment of specific agricultural potentials. In the U.S. the amount of data available is stupendous; there is danger of being overwhelmed by descriptive, statistical and analytical data. The National Agricultural Library, Beltsville, Maryland, is probably the finest resource of this kind ever assembled. Further, persons, not only at Beltsville, have devoted their lives to studying agricultural capabilities of particular regions. In addition, surprisingly, some underdeveloped countries have excellent data of their own and this can be the most significant data of all. Almost all countries certainly have the capacity to assess data and relate it to national peculiarities.

Also, NASA satellites already provide significant global agricultural data. The present generation of satellite pictures reveals in remarkable detail most of the planet's major land and water forms. They show, too, not only the scale and form of agricultural development but contrasts, usually on national bases, between agricultures. Soon new techniques will develop whereby definitioning will become even more pronounced. For the first time we may look at humanity as a whole and see what humanity is doing. This is not only a memorable achievement; it is pregnant with possibilities for agriculture. In terms of assessing potentials, it is one thing to offer a long compendium of statistics. It is another, however, to relate those statistics in graphic presentations. Indeed, soon the statistics themselves may be drawn from the graphics offered by new and more definitive pictures. Most important of all, everyone — peasant and president — can understand the pictures. It is difficult to argue with their contents. Thus, in the near future, we will at last see ourselves, what we have done and what we might do.

It will soon become possible to compile a planetary agricultural atlas based on NASA photography. Together, pictures, statistics, narrative and analysis could make available to all what has been done and what the planet might do agriculturally. Cost would be relatively small — a tiny fraction of what has been spent on agricultural aid to date. It should be especially important to the young of the Third World. Probably no single factor would be more important in purveying the truth about agriculture as it is and as it might be, as would this one volume.[1] It would probably become the most revolutionizing document published in this century.

The developed world is also the repository of much agricultural research; the problem again is that of being overwhelmed by quantity. But research has its own peculiarities. A great deal of it has little practical value; some might have been better left undone. Such wastage, however, is a necessary aspect of research. It is unavoidable and must be accepted as part of the research process. This was recognized a long time ago!

> **It seems to be a necessary condition of human science, that we should learn many useless things, in order to become acquainted with those which are of service.... There can be no greater impediment to the progress of science than a perpetual and anxious reference at every step to palpable utility.... Nor is it to be forgotten, that trivial and apparently useless acquisitions are often necessary preparatives to important discoveries.[2]**

This prescient passage sounds a warning. In practical agriculture one cannot afford to become swamped by research detail, much of it irrelevant. For example, long periods of experiment in the field, despite human impatience to attain aspirations, is one constraint to be observed. The research available must be distilled. Here the computer becomes the critical agent. Already the U.S. Department of Agriculture has computerized much of its agricultural data and more must yet be done. What is required is to make readily available, through computer recall, regional data as to what can be done and what potentially might be done for husbandry and new relationships of plants and animals to these regions. The primary purpose is not that of presenting data, important as that may be. It is to stir the imagination, to open the eye, to move, as Plato said, towards "correctness of the glance." In addition to basic research, exactly the same argument can be made for development or applied research and for techniques. This process, where the computer is king, could be related to the NASA-based agricultural atlas in an ongoing cycle of stimulating minds through truth as to agricultural potential. Until this is done we cannot set goals.

Relevation of truth is essential to seeking a course of action that has any chance of succeeding. But before action can begin, goals must be set. Probably no aspect of development work has been more poorly done than the setting of goals. There is good reason for this. Usually goal-setting is done by politicians. As we have noted, politicians do not always see agricultural development as an exercise in human development so much as they do in terms of aiding a political process, if not an individual politician. Too, we have noted ideology. Here goal setting can become very confused. There is sometimes an ideological rhetoric surrounding goal setting which obfuscates how those goals relate to individuals who must do the work. In the ultimate, most peasants are not interested in becoming new men, being harbingers of a new society or even representatives of the future. Peasants tend to be practical people. They are essentially interested in how much they are going to get, and when.

For any new development process to have a chance it must deal in tangibles and relate to people's interests. It is possible to activate enormous energies, as has been the case in the Soviet Union and PRC, by the initial change from the "old" to the "new." But once the change formally occurs difficulties arise. Now what? The question can be answered only by offering simple, tangible goals which relate quite personally to individuals.

One can begin with a general goal, say, to double the yield per unit area and to bring to cultivation a given area of new lands along with new crops, animals and techniques. But unless the state has decided to adopt a coercive posture, this general goal must be related to individual welfare. In deciding to give aid in any form to any program, this should be the primary factor. The reason is more than humanitarianism; it is confirmation of what aid is really about. It is also utilitarian. Without coercion, there will be no significant long-term development unless it is concerned with bringing tangible, recognizable benefits to individuals.

So the sponsor of aid carries heavy responsibilities within the recipient country.

It may also call for significant change in what increasingly is becoming a Western cultural attitude: that all should be equal. Here it will be a sponsor who must make the cultural change.

> **The rapid economic advance that we have come to expect seems in a large measure to be the result of...inequality and to be impossible without it. Progress...cannot proceed on a uniform front but must take place in echelon fashion, with some far ahead of the rest.[3]**

Hayek, in his mighty work, *The Constitution of Liberty,* goes further and makes another relevant observation on this issue of equality, especially as it concerns development:

> **This conception that all should be allowed to try has been largely replaced by the altogether different conception that all must be assured an equal start and the same prospects. This means little less than that the government, instead of providing the same circumstances for all, should aim at controlling all conditions relevant to a particular individual's prospects and so adjust them to his capacities as to assure him of the same prospects as everybody else. Such deliberate adaptation of opportunities to individual aims and capacities would, of course, be the opposite of freedom. Nor could it be justified as a means of making the best use of all available knowledge except on the assumption that government knows best how individual capacities can be used.[4]**

Here Hayek observes that instead of human capital burgeoning, it is curtailed.

Peasants have few illusions. For development to become a long-term prospect, the critical element becomes that of fair shares according to individual effort. Unless it can be shown that an individual farmer will get a fair share based upon his qualitative and quantitative input, the external donor would be wise to leave any project alone, for he will otherwise surely subsidize failure. Responsibility for insuring fair shares cannot be that of the external donor. To accept such responsibility is to become directly involved in governance of an alien society and there are no half measures here. One is totally involved, or one is not. If the recipient government cannot give viable proof that it can adhere to Hayek's admonitions, then that government is not fit to sponsor aid for its own people.

But there are more important features of goal-setting. Goals are meaningless slogans and misleading rhetoric unless they motivate the development of people. From this, and only from this, comes development of everything else, with more and better food and fiber the end result. There is danger, however, that NASA pictures, computers, and marvels of research become goals themselves. Goal setters become intoxicated by the fine words in which goals are often stated. Goals become a source of immense gratification to those who formulate them, instead of being the stimulus of those who must work in the field.

The humble peasant must understand the goals if these dangers are to be avoided. Here is a challenge. Peasants are not interested in rhetoric. Sometimes when they hear and see rhetoric spilling from political lips, they laugh — when the politician has gone. Peasants are as intelligent and practical as the rest of us. They are the first to know when something cannot work. The peasant in China, long before the Party hierarchy, knew that the Great Leap Forward had failed and that China stood on the threshold of unprecedented disaster. Thus besides being concise, simple, direct, and related specifically to individual needs and aspirations, goals must above all be practical and attainable. Establishing such goals is a challenge.

From goals arise blueprints for action. Once goals are set, action in its proper construct ought to become quite clear. It is, however, usually quite complex. In the past, the bane of development has been blue-prints which were essentially geared to the single factor theory. This led to sad results. The irrigation specialist, for example, whether local or foreign, learned that he was quite powerless, and ill-equipped, to handle local politics or indigenous cultures. In other contexts other specialists are similarly handicapped. Then there were projects that had no blueprints at all because no specific goals had been set. Here a kind of random chaos reigned. Each person associated with the project more or less did as he pleased. In this situation one sees a variety of specialties, all busily at work in a pointless kaleidoscope of variation.

The agricultural blueprint is a critical step in development. If broadly experienced, a Westerner can give significant help in its preparation. Very often specialist help is essential. But the specialties must be related; here is the key. For instance, a treatise on coffee trees must be related to transport, brokerage and international marketing. Another key is relating the agricultural plan to the overall national situation, to its international position, its ideology if such exists, its politics, bureaucracy and culture.

Here the Westerner is confronted by significant problems. Does he have the knowledge to assess? Is he being told the truth? Can adequate diplomacy be exercised to have unpopular action accepted? Will promises be kept? Is the government stable? How stable is the currency? Does the leadership abuse power? These are some of the questions.

Such problems are significant, but not insurmountable if the right persons from the west are involved. It would be presumptuous to write a prospectus for the kind of Westerner needed. Much is obvious. Acquired skills with experience, even wider knowledge, cultural empathy, an apolitical and a-ideological attitude, patience, a low-profile ego, all are needed. An especial quality is the ability to relate components into a coherent, comprehensive system. These all would be obvious requirements. The West does not have many people with these attributes. Certainly if farmers are required, most farmers who have these qualities are busy farming. In this regard we must conclude that in-country operations are severely limited if they must depend on widespread Western help. But the interrelationship of elements is the key to it all.

Rarely, if ever, can one single act or single material factor solve even seemingly simple issues. In complex issues, such as planetary agriculture, we see a matrix of problems. This calls for a matrix of remedies. Thus if, say, Bangladesh were to reduce its population by half, there might be some alleviation of its food problem, but no solution. The single factor theorist assumes that Bangladesh's food production would remain the same and thus there would be more for all. The fact is that Bangladesh's food production is directly tied to human labor. Reduction of the human labor output, with nothing else being done, would mean a near proportional drop in production. We could end up with about the same proportion of impoverished people.

As another example, it is true that more fertilizer and better seed strains are not only essential but critical to enhanced production of food and fiber. However, the thrusting of these elements upon a traditional agriculture without other actions being taken is unlikely to be effective. Such simplistic approaches may actually be counter-productive.

In each specific case, many elements will be involved in both cause and solution. What is needed is less a universal prescription than a series of interrelated parochial actions. These actions will involve less the didactic following of exact prescriptions than they will a series of compromises. There has to be a particular kind of compromise, a rigorous compromise. Anyone can compromise, especially if one does not care too much about the outcome. One gives way on any number of elements and in the degree required to placate whatever is in opposition.

A rigorous compromise is substantively different. One sets goals and then most rigorously adjusts elements to attain these goals. Here one hones each element to the sharpest edge one can and then adjusts honed elements to each other. One tests elements at variance one with the other and adjusts and adjusts to meet the requirements set by the goal.

Because pesticide-laden runoff poisons fish in streams, one does not either ban pesticides or say that fish do not matter. Rather, one seeks a rigorous compromise between restraining the effects of pesticides and preserving or enhancing aquatic life. And it will be a compromise. In agriculture one does not have the luxury of dealing in absolutes. The symbiosis itself is and always be a series of compromises. The question is the rigor of that compromise.

One disparity to the prescription must be noted. There are some single factor propositions that can theoretically produce striking results. One is fertilizer on lands that have not been fertilized before. Often increases of up to four times can be attained immediately. Allied with this possibility, a recent updating of the value of pesticides in non-pesticide-using countries noted that now up to half of all food production is obliterated by disease and primarily by insects and rodents.[5] Again we sound the usual caution. If a farmer is growing enough for himself and family, he is unlikely to produce a market surplus by using fertilizer and pesticides if the price of the surplus is fixed below cost of production. The American farmer reacts in a similar fashion to depressed market prices under the open market formula. When he felt that the price of fertilizer was too high relative to the price of crops, the American farmer has often cut back on fertilizer use.[6] Thus even when spectacular results can potentially be achieved by single factor means, the single factor must still be related to others. However, we should not overlook any potential, especially one so significant and so theoretically easy to attain as in the examples noted.

Another key element in the blueprint is to refer again to the concept of "mix." Technologies and other sophistications may have been selected which were practical in a particular situation and which could be related to local capabilities. A Westerner is tempted to advocate what he knows. He is reluctant to break up what he considers to have been a winning combination for him. He is often suspicious of retaining too much of the old which so often has been a losing combination. Again, devising a "mix" calls for an extraordinary degree of experience, to say nothing of paying the closest attention to local opinion. Only a few Westerners can even conceive of the concept of "mix," or are willing to bend their imagination in this direction. Thus we face another limitation in devising blueprints with Western help.

To date, the success of aid from the developed to the underdeveloped has not been notable, least of all when one considers capital investments made. The following table indicates the scale of overall aid to date.

PLANETARY AID TO AGRICULTURE FROM DEVELOPED COUNTRIES: IN MILLIONS US$ NET OUTFLOW[7]

	1970	1971	1972	1973	1974	1975	1976
Official Development Assistance	6,712	7,612	8,439	9,234	11,292	13,385	13,656
Other Official	1,111	1,242	1,551	2,403	2,254	3,024	3,305
Private Capital	6,258	7,271	7,893	10,786	11,373	22,428	22,186
Totals	14,080	16,125	17,884	22,434	24,919	38,837	49,147

Incorporated in part in the above is an accumulated debt of about $180 billion. This current indebtedness of underdeveloped countries, in all forms, agriculture and otherwise, will start to fall due mainly in the 1980's.

The huge debt hangs like a pall over the entire aid process and especially over the future of aid, agriculture or otherwise. It has to be dealt with.

Some suggestions in that regard will be made shortly. It also offers some lessons.

Developed and underdeveloped nations together generally have failed to stimulate agricultural production. The amount of capital involved probably could have been doubled, quadrupled, multiplied by ten, and results would have been much the same. The problem has not been money. The problem has been people and what they have done or have failed to do with these gigantic resources. It must also be remembered that in addition to donations, far greater capital investments than the monies indicated above have been generated and invested by the underdeveloped countries themselves. That, too, must be added to this assessment and taken into account in this declaration of failure.

This book has put forward general reasons for the failure. We must explore some specifics. We must ask how capital has been used in the past. How might it be used in the future? This is necessary for two reasons. First, capital is important. It is one essential lubricant to make any development move. Second, the underdeveloped world is demanding not only more capital but some alleviation of their enormous debt load. The developed world will probably respond to this in a simple but ancient way. Historically it has been common for wealthy nations to buy off an impoverished mass when it unceasingly shouts at the city wall. This solves nothing. Let us examine in some detail the process of transferring money from developed to underdeveloped.

In this assessment of the use of capital it is necessary to discriminate between gifts, loans and sadly, something that falls somewhere in between and is neither one nor the other. There are problems in giving away money even to people one knows and likes. The difficulty centers around the "strings" one puts upon one's gifts. Sovereign nations resent "strings," probably more today than at any other time. In fact some nations have refused point blank to accept gifts with "strings" which they felt represented subservience — "neo-colonialism."

Also, fron the donor's point of view, a gift with "strings" presents difficulties which under analysis make such gifts farcical. One gives money for a particular purpose. If conditions are not observed, what does one do? One can hardly send in an army to take the money back. One does not give again, under these conditions. Thus giving, especially with "strings," can create another block to development. By giving and failing, one ceases to give.

There is one process of giving that has not been tried although it probably offers few advantages. One could escrow a gift in the donee's name and release funds at a prescribed rate in accordance with successful performance. This would call for materially-agreed-upon criteria for judging performance. It would call for-on-the-spot evaluation by representatives of the donor. In the event of nonperformance it would call for an arbitrary decision to stop payment by the donor. In the way that national sovereignty is judged today, this would pose severe infringements on such sovereignty. It could create a very stressful situation for both parties. One can foresee occasions where such giving would create resentments and frustrations without especially guaranteeing any furtherance of result.

Another alternative, to give funds without any strings, has been the method most favorably viewed by recipients. This method also poses problems. If one does not really care how capital is disposed of, one can give and forget. It has a nice ring to it. Some Americans in particular often see this practice as possessing strong moral overtones. The morality of it is even more fervently believed in where a corresponding belief is prevalent: that poor, small countries are somehow more moral than rich, big countries. Be that as it may, there are three things wrong with gifts without strings, and maybe a fourth, depending upon one's point of view.

First, all nations are comprised of people, and some people — especially within ruling hierarchies — are not totally virtuous. Sometimes a gift without strings attached offers to the recipient the equivalent of a little gold mine wherein the smart, quick and on-the-spot privileged can extract such a bonanza that the wonder is that any stringless gifts were ever spent on legitimate purposes. The second factor is that the history of uncontrolled gifts to date tends, at least in part, to bear out the first contention.

The third fact is more important, more likely to eventuate, and more complex. What normally happens when stringless gifts are abused is that the malefactors are discovered, or at least suspicions are raised. It may be a foreign reporter, a visiting Congressman or even an embassy that first discovers that funds are being milked. Action quickly follows, action as deplorable as it is understandable. A bevy of inspectors, accountants and auditors appear from the donor country. They check, pry and spy. The malefactors attempt to checkmate and usually succeed. In the end the donor seems to come out on top because after all he controls the purse strings. Colonial overlords never played such a heavy-handed role with their wards as does a donor who originally came to the scene as the antithesis of colonial overlord: a true bearer of gifts. Misuse of gifts has maneuvered the donor unwittingly into an unpleasant and unsought role.

This writer well remembers a Laotian friend, a most courtly man, well known for his ability to siphon off, for himself and family, large percentages of U.S. aid funds. Ultimately his skills drew official American censure. "Ah," he would sigh, "never did the French when they were our colonial masters, breathe down our necks like these Americans!" Again the environment engendered by this kind of giving is hardly one to stimulate development of human capital. Yet we note that funds filched from donor nations have often been used to remarkable advantage by individual entrepreneurs. Sometimes the factory so financed was more attuned to the recipient country's needs, and was certainly built faster, than if the funds had gone through channels.

The fourth objection to gifts, which all might not agree to, is that they seem to erode national will and even morals. It is an ambiguous position when one finds oneself trumpeting nationalist slogans and flaunting perhaps a new independence and yet finds oneself the recipiet of money donated by a large and powerful benefactor. Most outsiders feel that behind the scenes there are strings, even if this is not so. People feel beholden. Even though nothing may have been demanded in return, people feel that demands could be made at any time. It builds an unhealthy relationship. What would the state of Anglo-American relations be today if those extraordinarily large amounts of British capital which came to the U.S. in the 19th and early 20th Centuries had been government-to-government gifts by England, rather than loans focused upon straight investment by private groups?

What would have happened if through American malfeasance, supervision by Englishmen had been imposed? Suppose the English supervisors had been 19th Century secular missionaries, full of righteousness, armed with English power but determined to lead these wayward American bankers, industrialists and railroaders back into the mainstream of Victorian morality and righteousness? Such interference, one imagines, would have strained American goodwill and probably would have left behind a legacy of ill will. Such actions might have assuaged English guilt (healthily absent) toward their late colonial wards, but it would have set American resentment aboil.

Much of what has been said about gifts of capital can also be said about gifts of food. There is some advocacy in the United States that the U.S. should grow even more grain and give large quantities (no one ever says how much) to the planet's needy. Even if the U.S. had capacity to feed the needy, which it does not, the process would create all of the difficulties noted as regards capital. It would do something else much more serious. We have already noticed that in certain circumstances U.S.P.L. 480 has acted as a disincentive to farming itself. One can imagine quite dreadful scenarios of societies waiting for their Kansas wheat. If one is to be given food, why grow it? Such "gifts" as an additional price depressant in already depressed markets would also be an effect. Moreover, the indigenous politicians would have a field day in manipulating their own markets while at the same time they could accord even less priority to agricultural development.

If one does not wish to give, one can lend. Conceptually, lending offers a particular advantage. Always in short supply and difficult to create, capital can be conserved by lending. Assuming that capital has been created in the first place, it is lent, utilized and then returned by the borrower to the lender. It can then be lent again. It is an ancient process and dates back at least to Sumerian times. Lending and borrowing is the heart-beat of modern commercial enterprise. When lending is under private auspices, that is, when there are as many lenders and borrowers as supply and demand allocates, the process is enormously flexible in its response to particularisms. It is adjustable to individual needs and becomes highly stimulatory to economic growth. Rather than viewing loans as government-controlled enterprises, which most are in the Third World, loans instead ought to be viewed from the point of view of the one who ought to be the end user, the farmer. Some loans never reach the farmer, but the only basis for judgment on loans worth bothering about is to assume they do. It is the farmer who develops, and how he may be treated becomes paramount.

A loan is a very special device and has "natural" laws pertaining to its use which, if neglected, negate its value. First, the loan should stimulate growth of a project, not stultify growth. This means that interest rates and general repayment terms must be fair and repayment should adhere to economic principles.[8] When a loan is seen to be fair, a significant step has been taken in development of human capital. Fairness breeds optimism. It engenders good management principles because the managers can see that if he does things right he becomes a beneficiary of a loan rather than a slave to the lender. In the latter instance "management" normally consists of preventing the loan from being called. In the first instance the effort is directed toward constructing a future. In the second instance the effort is to obstruct disaster. The sense of fairness, a sense that the loan is manageable, is entirely a human factor. Without it there can be no human development; and without the essential human input of energy, skills and creativity there will be no agricultural development.

A second critical feature of lending is that loans should never be made to any project where there is reasonable doubt that the project can support repayment. This factor is as vital as it is obvious. It calls for a most critical evaluation in the first instance of the entire project. In this evaluation the human aspect comes first. Is the borrower a capable farmer relative to the project proposed? Does he know how to use the new equipment if such be included? Have all other resources necessary to achieve success been mobilized? Above all does the farmer clearly understand the terms of the loan and his responsibilities thereto?

The lender then must look wider afield. Can the product be marketed? Are transportation and processing facilities available? What are the market forces at home and abroad? What human resources may effect marketing? The lender also must assess the political and bureaucratic forces at work. If extortion and bribery are paid as the *modus operandi* these must be evaluated in the context of an individual's relationship to this situation. If, for example, he has to pay off an official before he can move his product, this is an extra cost to be added to the regular cost of production. If the government fixes prices for crops, as is so widely the case, then a new condition arises calling for yet another kind of evaluation. If the price-fixing government conducts such activity for political purposes, to provide "cheap" food for the urban population, then probably the making of effective loans becomes impossible.

If prices are fixed with some regard for production costs and the economic forces of supply and demand, perhaps some constructive approach can be made. Also important is how much it costs individual farmers to obtain their share of the loan. In the Third World, governments as lending agencies are slow, inefficient and arbitrary. Slowness is serious. A crop cycle comes and goes, not in accordance with bureaucratic process, but according to nature which is at the same time variable, but quite inflexible. Does the farmer have to pay bribes at the lending table? How much time does he have to spend at a bureaucrat's window? Does he have to travel far? We must remember that most farmers do not have transportation, but must walk. This is a net cost. In most instances state agencies are not merely arbitrary and inefficient. They tend to favor the wealthier rather than the poorer farmer. The former has more leverage. He is a safer prospect. He knows more about farming.

The larger and wealthier farmer has better developed human resources. Instinctively, bureaucracy recognizes that fact. There must be a normalcy in lending. In most of the Third World abnormality is the mode.

We are seeking the practice of a tried and proven method of financing, the application of "bankers' criteria," which assures the viablility of a loan for both lender and borrower. Both have a mutual interest to be satisfied. Bankers' criteria do something else — they test projects. Experience would indicate this not only in economic terms, but also in terms of the viability of human resources available. If one must assuage ideological sensitivities, bankers' criteria can be called by some other name. The principle of testing against a set of fundamentals, however, remains. It holds for socialist societies as well as capitalist, for we are dealing with fundamental "laws" of nature.

If one does not apply bankers' criteria rigorously and make judgment based upon results of such evaluations, one should not lend. Lending under any lesser stricture offers all the bad features of lending and none of the obvious advantages. To lend to projects which have not been assessed using bankers' criteria is to play roulette with aid and development. The borrower has contracted to pay back, and now he probably cannot. Relations between lender and borrower under these circumstances are apt to deteriorate. If default becomes public, the borrower's ability to finance future projects is gravely impaired, even destroyed. If default of large sums of money were publicly announced, there could be adverse repercussions in the international money markets. We shall later ponder the fate of the ever increasing billions owed Western lenders by the Third World.

If one cannot apply bankers' criteria, one had better stay out of the aid process entirely. If one *must* offer aid, it would be better simply to give money and disregard how it is used — if one can. But no country is capable of such action. No country is rich enough, nor has the equanimity or irresponsibility, to allow capital it has created to be handled so dissolutely. Aid financing — lending or giving — has to have strings.

Before further delineating what those strings should be, we need to look at advantages inherent in a lending versus a giving program. The first value arising from lending relates to that much used term, nation-building. To borrow money and be held responsible for repayment is basic to fiscal responsibility. In terms of nation-building, fiscal responsibility is cardinal to a sense of worthiness in that nation, and also in social stability. Giving treats a nation as a ward. Nation-building calls instead for collective responsibility for the national destiny. To construct a national borrowing program means setting goals and objectives and establishing ways to meet these goals and objectives. Ends and means must be viable...something that will work. One should expect that such events lead to a sense of nationhood and, where goals are achieved, to national self-respect. Charity might not be so effective in attaining this end.

There are other assets to be had from borrowing and lending. A successful enterprise (which is what a loan operating in accordance with the tenets of bankers' criteria really is) builds up production competency. In the end, this is what enterprise is all about. Productive competency leads to self-assurance and in turn to an ability to convince lenders to lend again. If loans are repaid, the ability to lend yet again again occurs. In this situation capital can be "rolled over" indefinitely and real growth can ensue.

This has been a feature of all successful nation-building in the past; the concept is no less valid now. Instead, giving has the exact opposite effect. For recipients it has bred a sense of dependency which leads to concern about one's nationhood and, in turn, leads to dislike of the donor because he is the source of one's dependency. On the other hand, donors often want to be loved for their generosity. When they are not, they sniff ingratitude and their willingness to give declines. This has certainly been true of U.S. giving to date.

Other little-recognized features of giving have been the "hire and fire" and the "free purchase" syndromes. By contrast with lending, giving often has meant that a retinue of aid personnel accompanies the financing. Whether these foreigners were acceptable or unacceptable, whether they ultimately fitted into the society or remained alien to it, whether operationally effective or ineffective, they had to be endured by the recipient. Lending offers greater autonomy to the borrower. It is better if the recipient has the right to "hire and fire" no matter how arbitrary firing may sometimes be. Lending under bankers' criteria confers this right. In terms of hiring, the element of national autonomy is obviously the critical one. There are other advantages. Perhaps a Japanese firm can bring a factory to life more efficiently than can an American, even though the latter country may be the financier. Healthy competition never hurt any economic enterprise. In the end the American loser might gain by improving its performance.

In terms of "right to purchase" many aid funds have been linked to purchasing equipment and material from the donor country. This has discouraged competition, has often meant higher prices and even inappropriate equipment for a given task. Aid funds might be more effectively used if recipients could buy on the most favorable market and, under normal processes, decide themselves.

Lending obviously is the cheapest form of aid. At one end of the lending spectrum is repayment, which allows the same capital to be re-used unto generations yet unborn. At the other end is that lending, if in accordance with bankers' criteria, presupposes that the recipient country is going to do the job effectively itself, since "bankers' criteria" has established that fact or the loan would never have been made.

There are distinct limitations on borrowing. Borrowing can be only part of the financing development process. It can provide sufficient capital to finance only a small percentage of development costs. Capital in the quantities required simply does not exist in the world's capital markets. Most of the capital required must be created in-country. Estimates as to how much capital is needed vary widely, but no accurate figure can be determined until, at least, national goals have been set. The most commonly quoted figure on a planetary basis is approximately $700 billion or perhaps $30 billion per year over a period of 25 years. This represents a mere 1% of the *planetary* gross national product.

It does not matter whether the country concerned is socialist, capitalist or in between. Capital comes from creating surpluses and saving these for reinvestment. Most successful development programs in agriculture have succeeded on that basis.

The dramatic agricultural advances in Japan that post-dated the Meiji Reformation were financed internally. There were no gifts and no borrowings. In more recent times agricultural development in Taiwan received only small gifts from the U.S. The vast bulk of Taiwan's development was financed by internally-generated funds, as was in People's Republic of China. India, on the other hand, has received substantial capital as gifts and loans, yet India's productivity is still low. Thus lending and borrowing, even when expertly conducted and generously granted, cannot fund agricultural development *per se,* but only a relatively small part. Funding by and large can come only from a nation's own efforts.

There is yet another difficulty to be faced with the lending process. Quite tangible, it confronts us now. This is the establishing, as an agreed and ongoing process, of that critical element, bankers' criteria *per se.* This calls for a high degree of knowledge and skills on the part of both lender and borrower, skills which are not possessed in abundance anywhere. The lender must have intimate knowledge of a borrowing country's foreign relations, ideology, politics, bureaucracy, culture and social framework. In addition, the lender must be able to assess the viability of a particular project on its own merits and relate that viability, if such exists, to the wider matrix.

The borrower must be able to do much the same, but he must also demonstrate that he possesses a sponsorship group which can indeed sponsor development of the kind under consideration. Here all of the mechanics are involved. Dispensing of rural credit is a task that calls for banks and bankers in the field. Alas, one sees few loan officers talking to peasants, or peasants even knowing that such persons exist. What, too, of processing, transportation, marketing and pricing? If it is to be anything more than simple distribution of funds, lending and borrowing is a comprehensive, sophisticated business, but not so complicated and sophisticated that it is beyond the power of average people to accomplish.

In commercial societies, the process is routine. It proceeds, too, in the Third World, but only in the cities. For agricultural development to proceed through borrowing, the process must be extended into the countryside. This shortfall exists right now. Until it is rectified, loans will continue to be made upon a lender-to-government basis. What thereafter transpires in-country becomes the government's business.

In viewing or reading reports of projects so financed, two things stand out: First, the projects envisaged are usually well described and conceptually sound. Second, reports on the projects are good background material for preparing a loan agreement. However, they have little if any specific relationship to such an agreement. Banker's criteria are necessarily precise in detail as to how funds will be spent and repaid. Such is rarely, if ever, present in contemporary projects. We should not call these transactions loans. They are merely another method of distributing wealth: wealth that will ultimately be dissapated. The lenders are not bankers, but distributing agents.

In other words, all agricultural development aid should have strings. But the imposition of strings is difficult. To summarize, there are logical steps. First, the project must be accurately described, especially in terms of goals. This is straightforward and usually it is done rather well. Second, the project must be reduced to detailed financial dimensions. The task now becomes more difficult because much detailed research is required. The third step is to assess capabilities of the sponsoring agency. Such an agency is always necessary where one is endeavoring to modernize traditional agriculture. Has the sponsoring agency the capacity to Will it be allowed to work? Some of the skills required and conditions that must pertain have already been noted. It is these that any lending body must probe right at the working level.

This is probably the most vital task of all, and the most difficult. It is a judgment that must be made usually be persons from an alien culture. It calls for a wide, wide experience on the part of the judge or judges, not merely in agriculture and finance but in terms of governmental and cultural relations. There are not many people who can combine all of these qualities, but there are some.

The fourth point is diplomacy. One has to tell the truth. Any other course, even the diplomatic bending of words, leads to failure. The truth required consists of much of the material in this book. Our world of finite resources simply cannot afford to squander capital. To achieve the most effective use of capital is in the borrower's best interest. It presages success with all the particular and national benefits already noted. But the borrower has quite specific, exacting and skilled tasks to perform. How to say these words without appearing offensive, overbearing, patronizing and unduly mercenary? That is the task of diplomacy.

The Third World owes the developed world over $200 billion as of 1978. At least $40 billion of this enormous sum is held by private financial institutions. One can speculate that little of this money will be repaid. The ventures for which these funds were lent have for the most part been unsuccessful. Many of these were "soft" loans; some were made for "social purposes." If ever there was folly, here it is.

It is a contradition in terms to lend for social purposes. One can lend only for financial purposes. If one wishes to use money for social purposes that is a different issue from lending, or indeed, from financing generally. Probably under these circumstances one should give, write the money off at the inception and not be concerned with what happens afterward. This is not hard-hearted; it is an exercise in clarity of thought. Rather than being beset with guilt, a lender should remember that effective use of money in this instance means agricultural development. What could be more humane in today's world than that?

What of this enormous amount of capital borrowed by the Third World? The only way it can be repaid and re-used is for economic development to be effected. Some economic development is proceeding and some loans may be repaid. But regarding the bulk of the funds, there probably is little point in calling the loans when due, because permanent default serves nobody's ends. The loans could be written off but this is tantamount to writing off the help of the developed for the undeveloped. After a country has defaulted or has had a loan written off (which is the same thing) further loan funds are unlikely to be forthcoming.

Further, large-scale defaults would promote shock waves in international and national money markets from which no one could benefit. Thus we now face a dilemma for which the developed nations are as responsible as the underdeveloped. After all, it was the developed who made the loans, if that is what they must be called. If one seeks a constructive resolution, there now seems only one course of action open to both parties. First, the loans should be allowed to default technically and temporarily. If this does not happen — if they are re-negotiated in some form without defaults — no lessons will have been learned. The situation will repeat itself.

As soon as loans do default, a specific course of action should be determined, aimed at obtaining the original result. This would be accomplished by establishing strings *post facto*. The strings should have objectives outlined in this chapter, at least insofar as agriculture is concerned. In particular, creation of a viable sponsoring agency for agricultural development should be the primary requirement. It is also important that this loan reconstruction be made public in specific detail.

More capital will be required not only to replace that which has been dissipated but, on the assumption that development is proceeding, to fund new ventures. Some of this capital has to come from the Western sector of the developed world. The communist nations are essentially debtor nations. The Soviet Union and the East European states *in toto* owe the West over $50 billion as of 1978. Because capital must be drawn from the West, it must be largely private capital, for it is that sector which generates capital. It is unlikely that private capital will again be attracted to unerdeveloped regions without guarantees in terms of time frames for investment, guarantees against expropriation and, it is hoped, concrete indications of the viability of projects. In other words, renegotiation and further lending should occur only under the strictest bankers' criteria.

At this stage, in terms of the aid process from the developed to the underdeveloped, some interim lessons can be applied to all forms of aid, and agriculture is no exception. In the forefront, developed nations have hopelessly confused social and financial constructs, to the detriment of both. If one decides to pursue social objectives, these should be pursued by social means. It has been asserted, however, that this offers an extremely limited role to the developer. Person-to-person aid has been conspicuously unsuccessful and will continue to be, for it is based upon a false conception. Cultural divisions are simply too strong in their practical expression of differing mores, language and perceptions for such aid to work in the field. The developed can, however, offer advice at the conceptual level. The West in particular is in a constant process of agricultural research and development. While much of this will not fit into the underdeveloped world, some will. Even where it does not, such research and development stimulates the mind and opens new vistas to recipients. Research and development should be freely given to all who want it.

The West in particular, and also Japan, are constantly bringing to life new machines in support of agriculture. Although most of these are not suitable for the underdeveloped world, they do open up vistas. The small agricultural machines developed in Taiwan and Japan are part of a genre. In the U.S. pesticides may be sprayed from an airplane; in Taiwan two men with pesticide-laden back packs carry a perforated pipe between them and spray, alternately, by a hand pump affixed to the back pack. Both parties — U.S. pesticide sprayers and those in Taiwan — seek and obtain the same results. The West can also give valuable aid in designing and blueprinting projects, which it does very well. This kind of approach is often the only way to define a project and then transmit that project in practical terms over time and space. More importantly, the West has the technical expertise to devise development schemata even though these will always have to be adjusted to the local condition.

The developed can also lend personnel, not as field operators but as specialists giving their knowledge of a particular and needed specialty. This is probably a fairly rare requirement, but it does arise, and this service can be catalytic.

Under limited circumstances it might be possible to utilize aid terms from the developed world to add to the phalanx given above. But people matching appropriate experience with appropriate skills are few in number. To send anyone else, especially the inexperienced and unskilled who have a burning desire to do good, inhibits not only development, but instills the maximum personal trauma on all parties, while moving toward certain failure.

On the other hand, financial means should be used to pursue financial ends. To pursue social goals through financial means insures that these will not be attained, and insures financial loss as well. We can repeat a simple rule: the poorer the country the more precise and more stringent should be the financial "strings." A very poor country cannot afford either material or psychological failure. Sadly the opposite policy has been adopted and for good and kindly reasons. But good and kindly reasons do not enhance the productivity necessary to repay a loan, no matter how easy the terms. The poor simply have to have success. To increase an impoverished nations's GNP is a noble social goal. The record shows that loosely described schemes and imprecise financial constructs do not increase productivity. For example, one may devise a loan where the repayment schedule and repayment rate may be easier for the very impecunious (as opposed to the not so impecunious). The trouble with this procedure in the past has been that when this aspect of the loan was "soft," so was everything else. The opposite ought to be the case. To the very poor, loans should adhere to strict bankers' criteria. Let the terms be easy, but the criteria strict. The important thing is not to confuse financial and social goals and methods. Each has its own dynamic and each should be treated as an autonomous element.

There is a great advantage to the developed in making aid work. Mostly, especially in the West, humanitarian aspects are emphasized. These are worthy reasons for aid. There are equally important material reasons. U.S. trade imports and exports with the Developing Countries in 1977 was $111 billion. This is about the same as U.S. trade with all of Western Europe. It must be remembered that the Third World people with whom we do this significant volume of trade are but the tip of the iceberg. The Third World contains most of the world's peoples, some 3 to 3½ billion, and their economic base is very low. Western Europe contains a mere 250 million, and their economic base is high. Should even small increases in purchasing power come to the Third World, its impact will be explosive. There will be demand for all manner of manufactured goods and for food and fiber as well. It is not more mouths *per se* which places the real demand upon food and fiber supply, but affluence, the result of production. For example, the 3½ billion people of the Third World utilize about ¾ of a pound of fiber (mostly cotton) per annum; the 250 million people of Western Europe, about 30 lbs. of fiber. The potentiality is easily discerned and it is almost as great for food and manufactured goods.

Agricultural development is closely tied to industrial and other forms of development. Some have said that agricultural development furthers industrialization.

> **The so-called "industrial revolution" was not created by a few rather unimportant technical changes in the textile industry; it was the direct child of the agricultural revolution based upon turnips, clover, four course rotation, and livestock improvement which developed in the post-half of the eighteenth century. It is the turnip, not the spinning jenny, which is the father of industrial society.[9]**

Whoever is right in the argument, there is an obvious linkage. The orthodox view is that if a large and growing industrial population is to be fed, an agricultural surplus must be available to feed them. This seems to be absolutely essential at the beginning of industrialization. Later, should industrialization proceed vigorously, its products form wealth that can be used to buy food, for which there would be an increasing demand, from abroad if necessary. This was certainly the case for Great Britain from the early 19th Century to the present, and is currently the case with Japan.

Agricultural surpluses have generally been accorded the role of providing the seed capital upon which industrialization has begun. But both agricultural development and industrialization proceed because a certain environment has arisen which is conducive to human creativity. The real seed, as has been suggested so often in this book is the release of human abilities. The initial capital, therefore, is human and not material. In any event such development represents an expansion in demand and, in turn, an expansion in world trade.

Japan maintained a healthy population for centuries, until the immediate post-war years, basically upon rice, vegetables and a little fish, though in the first half of the 20th Century some of the rice came from Japan's colonies in Taiwan and Korea. In 1974, however, Japan imported $10.6 billion of U.S. goods, a total not exceeded by any other single customer. As part of this trade pattern an informal agreement now exists between the U.S. and Japan, whereby the latter is promised 14 million tons of grain for the next three years. Of the U.S. goods imported by Japan, more than one-third by value were agricultural products. In turn, the U.S. imported some $12.4 billion of goods from Japan.

One can observe a similar trend in Taiwan, which — tiny as it is — imports and exports to the U.S. over $7 billion. In contrast, India whose population outnumbers Taiwan by a ratio of about thirty to one, imported and exported about half a billion dollars by value. Indian trade with the U.S. was, in fact, less by value than that of Hong Kong, whose people very clearly are outnumbered by India's by about 150 to one. One must be careful, however, not to stretch these statistical comparisons too far. Hong Kong's position is quite different from India's. In 1978 Taiwan's trade with all of the world reached almost $24 billion. This was, in 1978, a little less than the PRC's total world trade even though the mainland outnumbers their fellow Chinese on Taiwan by a ratio of about 60 to one. Thus with general development, and in some instances with agricultural development being the seed, new vistas open in terms of trade relationships between developed and underdeveloped: the first step in development is giving incentive to individuals. Perhaps incentive to individuals is another way of defining human capital.

It might be asked why in this book no mention has been made of the international role of "transnational agri-business" or of "monopoly capitalism," to quote two contemporary cliches regarding Third World agriculture. Increasingly these elements are being assigned a devil-image role as interested parties find it increasingly difficult to continue charging that overpopulation and like factors cause agricultural poverty. One suspects that the emergence of Africa, where under-population is prevalent in many of the nutritionally deficient nations, is one reason for this embarrassing discovery. Despite dark mutterings, in terms of Third World agriculture, "transnational agri-business" and "monopoly capitalism" are not really very important, no matter how important they may be as regards such other resources as oil and some minerals.

In 1977, the world production of man's main food crops was 387 million metric tons of wheat, 370 million metric tons of rice, 349 million metric tons of corn, 293 million metric tons of potatoes, 174 million tons of cassava. The lesser crops comprise such less staple items as grapes, sugar cane, fruits and the like. It is the first category of grains and potatoes that provided the bulk of humanity's primary foodstuffs, and the vast bulk of this was consumed *in situ*. Only tiny percentages came to the world market. Some 11 million tons of rice, 3% of the world total, was exported. Had all of this export been manipulated in some nefarious fashion by "monopoly capitalism" one is left incredulous that somehow monopoly capitalism denied the remaining 97% left in home countries to people who grew the grain.

Much the same pattern applies to wheat, and especially to potatoes. In 1977-78 the U.S. which produced about 43% of the world's wheat exports, sold 31 million tons abroad, or about 8% of the world's total production. Of this, about 3.4 million tons was exported to the Soviet Union. The general opinion at the time was that it was the Socialists who manipulated the capitalists.

Regarding corn in 1977/78, of the 48 million tons, or some 73% of the world's export total sold by the U.S., most was purchased by Western Europe and Japan for animal feeding programs. No doubt one can find incidents where one rapacious Western company (U.S., if you like) drew unfair bargains with the governments of impoverished countries. Fruit is usually cited in this regard. The total banana world production in 1977 was 39 million tons, a mere 2.2% of total world grains and potatoes production. Again one wonders how, even if the total world banana crop were leveraged by some outside force, this in turn would significantly increase world food poverty.

We need to note some other facts regarding the Third World and its international commodity market generally. Many of the poorest and largest countries have little or no contact with the world food export markets. India and China, which together account for nearly half of the world's people and nearly half of the world's grains, export trivial amounts of foodstuffs. China imports much more than it exports, and often exports for purely political purposes. The Soviet Union is a large country which grows a large wheat crop, but it imports rather than exports. The main point is that "monopoly capitalism" even when assigned the worst roles by its critics, could have only the most limited effect on this sizable component of world foodstuffs. What does have an inhibiting impact on this massive segment of planetary agriculture is internal political and human arrangements, food priced below production cost, and other political devices such as taxes. Without doubt, no monopoly capitalist even in the days of colonial plantations has ever had such a negative impact on agricultural productivity as have indigenous governments.

Perhaps it is small countries to which the proponents of the "monopoly capitalist" theory refer. Often some small country producing a single crop, usually a fruit, is held up as an example of domination by "capitalism" or domination by a giant Western (usually U.S.) company which dictates not only in economic but also political terms. This has happened. The question is why in small nations and not somewhere else? Fiji, Taiwan and Hong Kong are small, but have strong international trading relations.

New Zealand, a tiny country, offers a prime example of successful trading. New Zealand has a mere 3 million people, yet is the world's highest per capita food exporter. In gross amounts, New Zealand exports more dairy products than any other country. These exports, including shipping arrangements, are totally controlled by the New Zealand Dairy Board. This Board is comprised of producer representatives working through producer cooperatives. As might be expected, these representatives are knowledgeable and dedicated to obtaining the best possible markets for their fellow farmers. They contract sales to private companies and governments throughout the world. Meat and wool, two other commodities of world significance, are handled similarly, without quite the degree of control exercised by the Dairy Board. These New Zealand Boards are strongly backed by politicians and a supportive bureaucracy. It is in everyone's interest that this be the case, because New Zealand lives by its exports. As anyone who has dealt with these Boards knows, they drive a hard bargain.

On the other hand, this kind of infrastructure, especially as it relates to producers, is simply not found in the Third World. It seems, in some instances, that governments do not care. In others, governments are unstable and more concerned about their security of tenure. In other instances governments are actively exploiting their own farmers.

Colonialism, as the last stage of imperialism, is often blamed for poor commodity trading performance. Why the British should be held responsible for, say, Burma's 1977 trade record, nearly 30 years after the British left that country, is a puzzle which cannot be tied to reality. It might also be noted that some former British colonies, New Zealand, Australia, Canada and the United States are today among the wealthier countries in the world. Of particular interest is the exceedingly high rice yields in *both* North and South Korea. These yields rival Taiwan and each is a direct legacy of Japanese colonialism in that peninsula.

The situation has probably been the opposite of that assumed by the devil image makers. Such intrusions made by the West in Third World agriculture have resulted in most of the significant advances — the introduction of new plants and animals, of new machine techniques, irrigation, husbandry and processing. Research and development results are freely available, for one cannot keep good husbandry a secret. Conversely, the intrusion of the West brought a massive decline in infant mortality, and this in turn stimulated high population growth rates. Similarly, much of the political rhetoric and practice of the West has been adopted by the Third World, and this rhetoric and practice give rise to behavior more typical of the hastier political activities of the West. Whatever else this may do it is hardly helpful to agriculture, as we have noted elsewhere.

Another relationship between the developed and the underdeveloped is that of food and fiber surpluses being used as leverages in foreign policy, thus having a deleterious effect upon development. Agri-power, as it is called in popular terms, can only be applied to one country, the U.S. Not only is the U.S. a superpower, but its food exports so exceed those of any other country that the two combined seem at first glance to offer a unique leverage. Can a superpower, which controls 75% of world food exports, have marketing power comparable to OPEC energy pricing power? This, as in the case with energy, could also lead to inflation, which would curb the transfer of wealth. One should beware of apparent similarities between food and oil. The situations are quite different.

The first difference is that U.S. food and fiber production involves nearly 2 million farmers who are autonomous decision makers. For a nation to use food, or any other commodity, as a foreign policy weapon, that country would at least have to impose some type of national food monopoly. Certainly export controls would be needed. All these steps are contrary to U.S. attitudes and conduct.

Of even greater significance, the impact of such programs on U.S. agricultural productivity could be incalculable. We have to remember those nearly 2 million autonomous decision makers. If wheat sales are controlled and wheat cannot be sold because withholding is part of foreign policy leverage, the farmer will, when he can, change to another crop or grow nothing for a season. The farmer will make the lives of his political representatives extremely uncomfortable.

A further difficulty in using food as a lever is that in agriculture there are great seasonal variations regulating supply-demand. This season there may be a shortage, but next year there may be a surplus. The leverage may not always be at hand. Such does not occur with oil. Japan, as one example, lived on rice, vegetables and fish for centuries. Today Japan produces a small surplus of rice, harvests a good deal of fish, and buys quantities of food abroad. But Japan could quite readily regress into a near sufficiency policy, albeit with a more spartan and less varied diet. The same could be said for People's Republic of China and the Soviet Union.

The United States enjoys some advantages from its vast productivity. There will be times — quite specific times — when food can help ease the U.S. through diplomatic negotiations.

> **Referring to Kissinger's Middle East peacemaking efforts, [Secretary of Agriculture, Earl] Butz says: 'the Russians could have blocked that agreement between Egypt and Israel when Henry was shuttling back and forth.' The reason they did not, the Secretary contends, is that they needed millions of tons of U.S. grain and 'they knew it was no time to be fooling around.' Asked if there was such a link between the Mideast Peace agreement and the Soviet grain deal, a high State Department official conversant with both negotiations concurs in one word: 'undoubtedly.'[10]**

Thus under certain circumstances and at particular times food could become a diplomatic factor. But this would be a bonus situation. The U.S. can hardly plan upon leverage from food as a diplomatic constant comparable to nuclear weapons.

As another example, if the availability of U.S. food can be more or less constant, other important effects could arise in diplomacy. Poland has been singled out. Poland, which hitherto has received 2 million tons of Soviet grain a year has been promised 2½ million tons of U.S. grain annually for five years. To what degree this might weaken Soviet control on its satellites is debatable, but it is potential leverage.

Food, too, becomes an important new factor in the overall economic strength of the U.S.; and economic strength is always a significant, perhaps decisive, element in negotiation. Food has to be added as a premier factor to leadership in the field of sophisticated technology, management skills at all levels, general finance and the fact that the U.S. offers the largest market to the world's goods. In some situations food could be a trade-off for scarce commodities, if need be returning to simple barter systems. It is a matter of matching two needs, and this is not always possible.

The real question about using food as a weapon, however, becomes a moral one. The U.S., as has OPEC with oil, could raise food export prices by fiat. The U.S. could, in times of shortage in a particular country, exert appreciable leverage by denying food or making it available under certain conditions. It is clear that people do not look upon food as they do oil. The price of oil can be dramatically inflated and it is accepted with a sigh. If the same position were adopted with food, there would be an indignant roar. Large numbers of people, including many Americans, suggest that surplus food should be given away. It is not our purpose to make introspections about double standards, but rather to note a situation that exists. This moral abhorrence of using food as a weapon therefore must be noted and must influence the way the U.S. adjusts its diplomacy.

Thus one might conclude that for practical as well as moral reasons food is far from being an "ultimate weapon." However, it has become a significant element in U.S. foreign policy and one which, if used properly, can further U.S. interests.

Our purpose is not to speculate on U.S. foreign policy even when this is related to food, but generally to survey relations between the developed and underdeveloped countries. Here again U.S. food surpluses might play a role. Situations could arise where U.S. food surpluses could act as incentives to increased food production by underdeveloped countries. We have noted that previous food gifts may have acted as a disincentive to agricultural development. A possibility would be to give food to a country as a supplement to a viable program of such development. Where, for example, a country is prepared to embark upon a program comparable to that of the Republic of China's Joint Commission on Rural Reconstruction, it might be a positive step to give food on decreasing scales as an initial phase of that program. As a necessary corollary, the U.S. should withhold food from countries which take no steps to sponsor vigorous agricultural development. Food surpluses might therefore offer another leverage, this time to aid development *per se.*

In this short survey of relations between developed and underdeveloped in terms of aiding the latter, the primary concern has been the state of mind of both parties toward development. Techniques are important, but these are passive devices awaiting activation by man. Further, we have a good grasp of how to set goals and blueprints, and how to implement those in a technological sense. We know what to do. The task is to get humans to do it.

First must come a sense of confidence that both the short and long-term future for humanity regarding food and fiber is good. The world *per se* has not rejected us yet. It offers us much bounty. Persons must know this. We must escape from our uninformed despair. We must also temper optimism. Results cannot be guaranteed. What can be guaranteed is the critical step of giving everyone a chance to optimize his own inherent abilities. The result of this latter point will upset at least one current trend most prevalent among the developed, that of seeking equality of result for all. The West in particular will have to adjust. Inequality is a condition of development. With development everyone is moving forward, albeit unequally. Inequality does not mean that some must be tossed aside. All can progress.

Nevertheless, the most important element is a new realization by the underdeveloped that basically they have to do the job. Once the underdeveloped realize not merely that they have to do the job themselves, but that they really can, the future becomes brighter. Perhaps the primary role of the developed is that in a humane way they might help bring about this state of mind. At that time, material aid from the developed becomes important. But the core of the problem remains and is not too much affected by new optimisms and the like. It stll rests with our five inhibitors, with the sixth lurking in the wings. Until these elements cease to inhibit and instead become progenitors, there will be little progress no matter how optimistic or even enthusiastic the ordinary citizen may become — developed or undeveloped.

FOOTNOTES

1 A beginning has been made although not yet in specific agricultural terms. See Nicholas M. Short, Paul D. Lowman, Jr., and Stanley C. Freden, *Mission to Earth: Landsat Views of the World.* (Washington, D.C., National Aeronautics and Space Administration, 1976). One can imagine the impact of say a comparison between the Mekong and Mississippi River valleys. These are similar in size and general topography. They are totally dissimilar in the agricultural productivity — and the question must be — why?

2 S. Bailey, *Essays on the Formation and Publication of Opinions* (London, 1821), Preface.

3 F.A. Hayek, *The Constitution of Liberty* (Chicago, Univ. of Chicago Press, 1960), pp. 42-43.

4 *Ibid.,* pp. 92-93.

5 American Academy for the Advancement of Sciences, 143rd Meeting, Denver, February 1977.

6 *Business Week,* "Why the Fertilizer Forecasts are Wrong" (April 19, 1976), pp. 132-34.

7 From UN Statistical Yearbook, 1975.

8 A "fair" loan bearing low interest rates has often had a peculiar fate in the Third World. This writer has observed that on some occasions peasants who were accustomed to paying thirty percent and more to money lenders were tardy in paying eight percent at a legitimate

bank. The peasant reasoning was that the interest rate was so small that a repayment of it was too insignificant a matter to bother about.

9 K.E. Boulding, quoted in Hayek, *op. cit,* p. 525.

10 *Business Week,* "U.S. Food Power, Ultimate Weapon in World Politics?" (December 1975), p. 56.

12 THE DETHRONEMENT OF POLITICS

The future looks stark. The inhibitors of human productivity are well entrenched. The centrist, regulatory state is now a near universal phenomenon. Differences between states, be they called democratic or totalitarian, is now only that of degree — the degree in the rate at which human liberties are eroded: degree in the nature of the coercion used. There will be no reversal of the trend unless ordinary people, the great majority of us, awaken.

We must demand nothing less than the dethronement of politics, that is the devolution of state apparatus as it is and a return of power over lifeways back to citizens. Humanity can no longer afford the varying processes of government as these are expressed in the contemporary world and described in this book.

Neither can the democracies escape censure. There, politics has become little more than bidding for votes through promises, promises which (to be fulfilled) must be redeemed in cash. But the cash must first be wrested from the taxpayers. Thus, those to whom the promises were made must pay. But it is worse than that!

Besides errant ethics there are practical consequences. We have noted the promises of cheap food in former chapters and how such promises have to result in disincentives to farmers. Thus production lags and instead of cheap food the people have expensive food accompanied by increasing inflation. Today US politicians promise "cheap" energy while they control the price of domestic oil and subsidize the price of the artificially aggrandized foreign oil. Domestic production languishes, inflation is boosted and the world's greatest economic unit is brought face to face with economic disaster. To repeat, we can no longer afford the state — and we shall face worse ahead.

How do we combat the state? The most prevalent form of reaction against a state is revolution. But revolutions almost invariably induce a worse tyranny than they sought to replace. The American Revolution was an exception. So was the "Glorious Revolution" of 1688-89 in England. It was also bloodless. The Revolution of 1688-89 essentially aimed at codifying certain civil liberties although it was the later American Bill of Rights which more completely codified these liberties and protected them by law. What we must now do is implement the equivalent in terms of economic liberties. What we have failed to realize (despite Adam Smith) is that civil liberties and economic liberties are synonymous. The former cannot exist long without the latter. Without economic liberty for the individual, codified under law and protected by an independent court, inevitably the state will abuse its economic power and inevitably some form of economic dictatorship will arise. Economic dictatorship by definition must in the end politically lapse into totalitarianism. Most of the planet is so afflicted. Most of the planet has yet to get even its Bill of Civil Rights, let alone a Charter of Economic Rights. More importantly, those nations not yet sorely afflicted are moving rapidly in the totalitarian direction. Even the U.S.A., the free-est country in the world and thus the most economically successful, slides rapidly towards economic dictatorship. Again, what can we do about it?

We must see what is happening. That is why this book was written. We need more books. We need to pursue enlightened reality which leads us to new attitudes toward the state. Then we must translate those attitudes into laws that control the state. These essentially come from an American perspective but they do enumerate certain principles that with variation apply to all states.

We must reaffirm an old truism. The state does not produce wealth. Wealth, whether it be cabbages or a work of art, is produced by individuals. Thus the state by law must be prevented from inhibiting the production of wealth. When legislation is proposed it must be preceded by an *independently* prepared economic impact statement as to how that legislation affects productivity, short and long term. We need the economic equivalent of the U.S. Supreme Court but working before the fact rather than after. This should be the basis of any proposal becoming law.

We must realize that in terms of the centralized management of an economy the state is incompetent. It is not that individual politicians or bureaucrats may not be competent or for that matter have high personal integrity. It is that a national economy is so complex that it cannot be managed by centralized controls. Such controls will produce stagnation, inflation and economic recession. Thus the charter of every bureaucracy as the operating arm of the political process needs to be reviewed with two objectives in mind. Is the bureaucracy necessary? If it is, how can it be made a service installation and denied regulatory powers? Where bureaucracy is not necessary it should be abolished in totality.

The bureaucracy should be decentralized into units small enough to function only as service agencies and these should be geographically dispersed. There would be obvious exceptions such as foreign affairs or defense.

The state must also be controlled in how much it can tax. Governance rather than becoming the fiefdom of a new, professional ruling caste must come back into the domain of the citizen at large. Each culture on this planet must do it in its own style because culturally we are different. But we should all welcome and learn from the diversity of method. The principle however, that of a Charter of Economic Liberties, needs be universal.

If humanity cannot attain this condition through conscious action, a historic change faces the planet for the inhibitors are pushing productivity towards a breakpoint. It is not too difficult to envisage what is most likely to happen. In those states that eschew force as a prime mover in government the trend is already discernible. It is a return to a medieval type of society. The poor must remain poor. The no growth philosophy and its attendant inhibiting political and bureaucratic actions ensures this much. Those who produce must remain producing sufficient to provide the state with sufficent wealth to pay off the poor and support the new governing class of politicians, bureaucrats, consultants, experts in this and that, foundations and what have you? It is a ceasing of human development. It is everyman in his place with his present rewards and obligations. It is a stasis — a twenty-first century medievalism.

At the other end of the spectrum, in those states that embrace force as the coercive method, we shall see greater use of force. Cambodia, 1975-78, represented a turning point. The massive slaughter of citizens in that land was not an unfortunate phenomenon reminiscent of other totalitarian ideologies in action. It was the logical extension of ideology that embraces the use of force by the state. The regime recognized that it was no longer enough to kill only dissenters. It was also necessary to kill those whom the state adjudged to be ineducable — those who because of age or background could not be turned into "new men."

Thus the dethronement of politics is more than that of allowing a citizen economic freedoms. Dethronement is essential to the process of civilization itself for the civilizing process continues only through diversity of human creativity and work. To reach optimums in creativity and work liberty is the essence.

This was a book about farms and farmers but perhaps it ended up as a parable. The story of the individual as potential creator facing the omnipotent and destructive state stands for all seasons. In real terms in this book we saw the state in its bickering and quarreling with other states and by these means inhibiting individual opportunity. We saw the state in its ideological mode physically crushing the individual. We saw the state committing political acts to preserve itself and its privileges at the expense of the individual. We saw the state in its bureaucratic arm bumbling and stumbling as it sought to do what it could not do and ending by becoming a center for extortion and bribery. Only in terms of culture as inhibitor did the state appear to be free of indictment. But here too it is the state that so often denies or limits the free exchange of ideas and methods that allows one culture to learn from another. The real enemy of the wealth of nations has been identified. The state and its politics must be dethroned: that is, cease to rule and instead serve. And what a future awaits us when the task is accomplished!

It is a future that requires both vision and the prosaic. We as farmers will banish the demon of material impoverishment which for so long has sat on humanity's shoulder. Others among us will perform less prosaic deeds and will realize new miracles that at first glance seem more important and creative than growing a field of potatoes. Some of us may even venture among the stars. If we do, we must remember to take with us our cow.

BIBLIOGRAPHY

1 James G. Abegglen, *The Japanese Factory* (Illinois, Free Press, 1958).

2 Economic Research Service, U.S. Department of Agriculture, *World Fertilizer Review and Prospects to 1980/81,* Foreign Agricultural Economic Report No. 115.

3 Economic Research Service, U.S. Department of Agriculture, *World Fertilizer Situation 1975, 1976 and 1980.* WAS 5 — Supplement, October 1974.

4 L.H. Bailey, Manual of Cultivated Plants (Rev. Ed.) (New York, MacMillan, 1949).

5 S. Bailey, *Essays on the Formation and Publication of Opinions* (London, 1821).

6 A. Doak Barnett, *Communist China: The Early Years 1949-55* (New York, Frederick A. Praeger, 1964).

7 A. Doak Barnett, *Uncertain Passage, China's Transition to the Post-Mao Era* (Washington, D.C., The Brookings Institution, 1974).

8 Harold J. Barnett, "The Myth of Our Vanishing Resources," *Readings in Human Population Ecology,* Ed. Wayne H. Davis (New Jersey, Prentice Hall, Inc., 1971).

9 P.T. Bauer, "Western Guilt and Third World Poverty," *Commentary* Vol. 61 (January, 1976).

10 P.T. Bauer and B.S. Yainey, "Against the New Economic Order," *Commentary* Vol. 63 (April, 1977).

11 Saul Bellow, *Mr. Sammler's Planet* (New York, Viking Press, 1969).

12 J.J. Benetiere, "Co-operative Strategies vis-a-vis Multinational, Agri-food Companies," *World Agriculture* (Paris, Vol. XXV, 1976).

13 *The Bhagavad Gita,* translated by S. Radhakrishnan (New York, Harper Bros., 1948).

14 Donald T. Bogue, *Principles of Demography* (New York, John Wiley & Sons, 1969).

15 G. Borgstrom, *The Hungry Planet* (New York, The MacMillan Co., 1965).

16 Francoise Bourliere, *Mammals of the World* (New York, Alfred A. Knopf, 1955).

17 Lester R. Brown, *Increasing World Food Output* (New York, Arno Press, 1976).

18 Nathan Buras, *Scientific Allocation of Water Resources* (New York, American Review Publishing Co., Inc., 1972).

19 James Burnham, *The Managerial Revolution* (Westport, Conn., Greenwood Press, 1972).

20 *Business Week,* "U.S. Food Power, Ultimate Weapon in World Politics?" (December, 1975).

21 *Business Week,* "Why Fertilizer Forecasts Are Wrong," (April 19, 1976).

22 E.H. Carr, *Michael Bakunin* (New York, Vintage Books, 1937).

23 A.M. Carr-Saunders, *World Population* (London, Frank Cass & Co., Ltd., 1964).

24 Winberg Chai, *The New Politics of Communist China* (Pacific Palisades, Calif., Good Year Publishing Co., 1972).

25 Nayan Chanda, "Hanoi Comes Down to Earth," *Far Eastern Economic Review* Vol. 95 (Feb., 1977).

26 "Pigs, Resistance to Collectivization," in *China News Analysis 1037* (Hong Kong, April 16, 1976).

27 Arthur C. Clarke, *The Lost Worlds of 2001* (New York, Signet Classic, 1972).

28 Arthur C. Clarke, *Report on Planet Three and Other Speculations* (New York, Signet Classic, 1972).

29 Marion Clawson, Editor, *Natural Resources and International Development* (Washington, D.C., The Johns Hopkins Press, 1965).

30 Norman Cohn, *Pursuit of the Millennium,* 2nd Edition (New York, Harper Torch Books, 1969).

31 Barry Commoner, *The Poverty of Power* (New York, Alfred A. Knopf, 1976).

32 Comptroller General of the U.S.A., "Disincentives to Agricultural Production in Developing Countries" (Washington, D.C., Nov. 26, 1975). [Report to the Congress.]

33 U.S. *Congressional Record*, "The Economics of Food in American Foreign Policy," 94th Congress (Senate), Sept., 1975, S15110.

34 R. Conquest, *The Great Terror* (London, MacMillan, 1968).

35 R. Conquest, *Power & Policy in the U.S.S.R.* (New York, Harper & Row, 1961).

36 Edward Crankshaw, *The Shadow of the Winter Palace* (New York, the Viking Press, 1976).

37 Joseph Darmstadter, *Energy in the World Economy: A Statistical Review of Trends in Output, Trade & Consumption Since 1925* (Washington, D.C., The Johns Hopkins Press, 1971).

38 Wayne H. Davis, Editor, *Readings in Human Population Ecology* (New Jersey, Prentice-Hall, Inc., 1971).

39 Wm. Theodore de Bary, Editor, *Sources of Chinese Tradition,* Vols. I & II (New York, Columbia University Press, 1960).

40 Douglas Diamond, *U.S. & U.S.S.R.: Selected Indicators of Agricultural Activity and Productivity* (Washington, D.C.: Conference on Soviet Agriculture, Kennan Institute, Nov. 16, 1976).

41 Victor D. DuBois, *Food Supply in Mali,* Vol. XVI, No. 1 of Field Staff Reports, West Africa (Hanover, N. H., American University Field Staff, 1975).

42 Anthony Dyson & Bernard Towers, General Editors, *China and the West: Mankind Evolving* (New York, Humanities Press, 1970).

43 Martin Ebon, *Lin Piao, The Life and Writing of China's New Ruler* (New York, Stein & Day, 1970).

44 Carl Eicher & Lawrence Wirt, Editors, *Agriculture in Economic Development* (New York, McGraw Hill, 1964).

45 T.S. Eliot, *Four Quartets,* Dry Salvages I (New York, Harcourt, Brace & World, 1963).

46 Jacque Ellul, *The Technological Society* (New York, Vintage Book, 1967).

47 Arghiri Emmanuel, "The Multinational Corporations and Inequality of Development," in *International Social Science Journal* (UNESCO, Vol. XXVIII, 1976).

48 U.S. Energy Research & Development Administration, *Agricultural & Industrial Process Heat Branch: Summary Report* (Washington, D.C., June 1976).

49 U.S. Energy Research & Development Administration, *Fossil Energy Program Report* (Washington, D.C., ERDA 76-10, 1976).

50 U.S. Energy Research & Development Administration, *Fuels from Biomass* (Washington, D.C., Nov. 1976).

51 U.S. Energy Research & Development Administration, *Solar Energy for Agriculture and Industrial Process Heat* (Washington, D.C., May, 1976).

52 Paul R. Erlich, *The Population Bomb* (New York, Ballantine Book, 1968).

53 Emil L. Fackenheim, *Encounters Between Judaism and Modern Philosophy* (New York, Basic Books, 1973).

54 John S. Furnivall, *Netherlands India: A Study of a Plural Economy* (Cambridge University Press, 1939).

55 Susan George, *How the Other Half Dies: The Real Reasons For World Hunger* (Montclair, Washington, Allanheld, Osmun & Co., 1977).

56 Edward Gibbon, *The Decline and Fall of the Roman Empire* Vol. I (Chicago: Encyclopedia Britannica, 1952).

57 John Gittings, *The World and China 1922-72* (New York, Harper & Row, 1974).

58 Michael H. Glantz (ed.), *The Politics of Natural Disaster: The Case of the Sahel Drought* (New York, Praeger Publishers, 1976).

59 Heather Goldman and Chandrashekar Ranade, *Food Consumption Behavior by Income Class in Rural and Urban Philippines* (Cornell University occasional paper No. 90, November, 1976).

60 Robert Greenleaf, *Servant Leadership* (Ramsey, N.J., Paulist Press, 1977).

61 F.A. Hayek, *The Constitution of Liberty* (Chicago, Univ. of Chicago Press, 1960).

62 F.A. Hayek, *Individualism and Economic Order* (Chicago, 1948).

63 F.A. Hayek, *Studies in Philosophy, Politics and Economics* (Chicago, Univ. of Chicago Press, 1967).

64 L.A. Heinail (ed.), *Hidden Waters in Arid Lands:* A Report of a Workshop on Groundwater Research Needs in Arid and Semi-arid Zones, Paris, Nov. 25, 1974 (Ottawa: International Development Research Center).

65 T.A. Heppenheimer, *Colonies in Space,* (Harrisburg, Pa., Stackpole Books, 1976).

66 Herman Hesse, *Journey to the East* (New York, Farrer, Straus & Giroux, 1956).

67 Fred Hirsch, *Social Limits to Growth* (Harvard University Press, 1976).

68 International Crops Research Institute for the Semi-Arid Tropics, *Annual Report 1974-75* (Hyderabad, 1-11-256, March 31, 1975).

69 Agency for International Development, *Techniques for Assessing Hydrological Potentials in Developing Countries,* TA/OST 73-17, (Washington, D.C., 1973).

70 Rachavan N. Iyer, *The Moral and Political Thought of Mahatma Gandhi* (New York, Oxford University Press, 1973).

71 Harry V. Jaffa, *The Crisis of the House Divided* (Seattle, Univ. of Washington Press, 1973).

72 D. Gale Johnson, *The 10th Five Year Plan, Agriculture and Prospects for Soviet American Trade,* Conference on Soviet Agriculture, Kennan Institute, Nov. 16, 1976.

73 D. Gale Johnson, *Theory and Practice of Soviet Collective Agriculture,* Conference on Soviet Agriculture, Kennan Institute, Nov. 16, 1976).

74 D. Gale Johnson, *World Food Problems and Prospects* (Washington, D.C., American Enterprise Institute for Public Policy Research, 1975).

75 Herman Kahn, William Brown and Leon Mandel, and the Staff of the Hudson Institute, *The Next 200 Years* (New York, Wm. Morrow & Co., 1976).

76 Robert G. Kaiser, *Russia: The People, the Power* (New York, Simon & Schuster Pocket Books, 1976).

77 A. Katsenelingoigen, "Colored Markets in the Soviet Union" in *Soviet Studies,* Vol. XXIX (Univ. of Glasgow, Jan. 1977).

78 Paul J. Kramer, *Plant and Soil Water Relationships* (New York, McGraw Hill, 1949).

79 Maurice M. Kelso, William E. Martin, Lawrence E. Mark, *Water Supplies and Economic Growth in an Arid Environment: An Arizona Case Study* (Tucson, The University of Arizona Press, 1973).

80 Edwin E. Kinter, "Statement on the Magnetic Fusion Program," Subcommittee on Fossil and Nuclear Energy Research, Energy Research and Development Administration (Washington, D.C., 1977).

81 Russell Kirk, *Enemies of the Permanent Things* (New York, Arlington House, 1969).

82 Hans H. Landsberg, Leonard L. Fischman, and Joseph L. Gisher, *Resources in America's Future* (Washington, D.C., The Johns Hopkins Press, 1963).

83 Joseph R. Levenson, *Confucian China and Its Modern Fate* (Berkeley, Univ. of California Press, 1968).

84 Moshe Lewin, *Society, State and Ideology During the First Five-Year Plan,* unpublished pending incorporation into volumes on Soviet social history.

85 Miriam and Ivan D. London, "Prostitution in Canton," *China News Analysis 1046* (Hong Kong, July 9, 1976).

86 Liu Shao-ch'i, *How To Be A Good Communist* (n.p., n.d., etc.).

87 Edward N. Luttwak, "Seeing China Plain," *Commentary* (December, 1976).

88 Norman Macrae, "The Coming Entrepreneurial Revolution," *The Economist* (December 25,1976).

89 Norman Macrae, "United States Can Keep Growing — and Lead — If It Wishes," *Smithsonian* (July, 1976).

90 Karl Marx, *Poverty or Philosophy* (London, M. Lawrence).

91 Carl Manzani, *The Wounded Earth* (Reading, Mass., Young-Scott Books, 1972).

92 Shakuntla Mehra, *Some Aspects of Labor Use in Indian Agriculture* (Cornell University Occasional Paper No. 88, June, 1976).

93 Marina Mehshikova, *The American Way in Agriculture and Its International Significance.* Paper prepared for 1976 Conference, "United States and the World," Smithsonian Institution, Washington, D.C., Sept. 27 — Oct. 1, 1976.

94 Frank C. Miller, *Knowledge and Power: Anthropology, Policy Research and the Green Revolution.* Unpublished paper, Univ. of Minnesota, August, 1975.

95 James R. Miller, *Models of Socialist Agriculture: The Soviet Case in Historical Perspective.* Conference on Soviet Agriculture, Kennan Institute, Nov. 16, 1976.

96 Norman N. Miller, "Journey in a Forgotten Land, Part II: Food and Drought, The Broader Picture, Ethiopia, Kenya," *Field and Staff Reports* Vol. XIX (Washington, D.C., American Univ., 1974).

97 Rufus E. Miller, Jr., Awakening From the American Dream (New York, Universe Books, 1976).

98 Frank E. Moss, *The Later Crisis* (New York, Praeger Publishers, 1967).

99 Malcolm Muggeridge, *The Green Stick: Chronicles of Wasted Time,* Vol. I (Glasgow, Fontana/Collins, 1975).

100 Raymond L. Nace, "Planning Water Supply for the Future," (Washington, D.C., Dept. of Interior, 1967).

101 Ronald C. Nairn, *UN Aid to Thailand: The New Colonialism?* (New Haven, Yale Univ. Press, 1966).

102 George N. Nash, *The Conservative Intellectual Movement in America* (New York, Basic Books Inc., 1970).

103 National Aeronautical & Space Administration, *Spinoff 1976, Technology Utilization Program Report* (Washington, D.C., April, 1976).

104 New Zealand Official Year Book 1974, Section 14, *Farming* (Wellington, Government Printer, 1975).

105 New Zealand Official Year Book 1974, Section 21, *Marketing* (Wellington, Government Printer, 1975).

106 Chiaki Nishiyama, *Pessimism in Spite of Omnipotence: Optimism in Spite of Limitations; Comment on the Report of the Club of Rome.* Paper presented to the Fellows, Woodrow Wilson International Center for Scholars, November, 1976.

107 Chiyaki Nishiyama, *The Price of Prosperity* (Hobart Paper, 1974).

108 Chiyaki Nishiyama, *The True Limits to Economic Growth and Progress.* Paper presented to the Fellows, Woodrow Wilson International Center for Scholars, November 30, 1976.

109 OECD Agricultural Policy Report, *Agricultural Policy in New Zealand* (Paris, 1974).

110 William and Paul Paddock, *Time of Famines: America and the World Food Crises* (Boston, Little Brown & Co., 1976).

111 G. Parthasarathy and Mihinder S. Mudahar, *Foodgrain Prices and Economic Growth* (Cornell University occasional paper No. 89, June, 1976).

112 Herbert Passin, *Society and Education in Japan* (Columbia Univ. Press, 1965).

113 John Passmore, *Man's Responsibility for Nature* (Old Working Survey, The Gresham Press, 1974).

114 Lewis J. Perl, "Ecology's Missing Price Tag," *The Wall Street Journal* (Aug. 10,1976).

115 Executive Office of the President, *The National Energy Plan* (Washington, D.C., U.S. Government Printing Office, 1977).

116 Jean Raspail, *The Camp of the Saints* (New York, Scribner's, 1975).

117 Edwin O. Reischauer and John K. Fairbank, *East Asia: The Great Tradition* (Boston, Houghton Mifflin Co., 1958).

118 Jack P. Rubel, A Report to the Secretary of the Interior, "Trust Territory Islands of the Pacific: Food, Fiber, Protein Action Proposal," (Phoenix, Arizona, May, 1974).

119 Katherine Scherman, *Daughter of Fire* (Boston, Little Brown & Co., 1976).

120 Arthur Schlesinger, Jr., "Laissez-faire Planning and Reality," *The Wall Street Journal,* July 30, 1975.

121 Peter H. Schuck, "National Economic Planning: Slogan Without Substance," *The Public Interest,* No. 45 (Fall, 1976).

122 E.F. Schumacher, *Small is Beautiful* (New York, Harper & Row, 1973).

123 Joseph A. Schumpeter, *The Theory of Economic Development* (Cambridge, Mass., Harvard Univ. Press, 1947).

124 Franz Schurmann, *Ideology and Organization in Communist China,* 2nd Ed., (Berkeley, Univ. of California Press, 1968).

125 *Scientific American,* Vol. 235, September, 1976.

126 James C. Scott, *The Moral Economy of the Peasant* (New Haven, Yale Univ. Press, 1976).

127 Fr. Joseph S. Sebes, S.J., "History of the Jesuits in the Old China Mission (17th & 18th Centuries): An Attempt at Cultural Accomodation." Publication pending.

128 U.S. Senate Committee on the Judiciary, *The Human Cost of Communism in China,* 92nd Congress, 1st Session, 1971.

129 Theodore Shabad, *China's Changing Map* (New York, Praeger, 1972).

130 David L. Sills, "The Environmental Movement and Its Critics," *Human Ecology,* Vol. 3, No. 1 (New York, Plenum Publishing Co., 1975).

131 Sithiporn Kridakara, *Rice Farming in Siam* (Bangkok, Suksit Siam, 1969).

132 J. Stanford Smith, "A Perspective on Regulation," *The Wall Street Journal,* June 14, 1976).

133 Earl of Stanhope, *Life of William Pitt,* Vol. 1 (New York, Amsterdam Press, 1970).

134 NARODONASELENIYE (Moscow: "Statistika," 173).

135 Benedict Stavis, *Making Green Revolutions, The Politics of Agricultural Development in China,* (Cornell University, Rural Development Monograph Series, 1974).

136 Benedict Stavis, *Rural Local Governance and Agricultural Development in Taiwan* (Cornell University, Rural Development Monograph Series, 1974).

137 Luis Taruc, *He Who Rides the Tiger: The Story of an Asian Guerrilla Leader* (London, Chapman, 1967).

138 Paul Tillich, *A Theology of Culture* (New York, Oxford Univ. Press, 1959).

139 David Keith Todd (ed.), *The Water Encyclopedia,* Printed for the Water Information Center, New York, 1970.

140 Leo Tolstoy, *How Much Land Does A Man Need,* in *Russian Stories and Legends,* trans., Louise and Aylmer Maude (New York, Pantheon Books, 1967).

141 Arnold Toynbee, *Mankind and Mother Earth* (New York, Oxford Univ. Press, 1976).

142 Arnold Toynbee, *A Study of History,* Vol. XII (London, Oxford Univ. Press, 1961).

143 William Tung, *Revolutionary China, A Personal Account* (New York, St. Martin's Press, 1973).

144 Colin Turnbull, *Mountain People* (New York, Simon & Schuster, 1972).

145 United Nations, Economic Commission for Asia and the Far East, *Economic Survey of Asia and the Far East* (E/CN.11/1047, March, 1972, Bangkok).

146 Richard L. Walker, *China Under Communism, The First Five Years* (New Haven, Yale Univ. Press, 1955).

147 Richard L. Walker, *The Continuing Struggle: Communist China and the Free World* (New York, Athene Press, 1958).

148 James A. Weber, *Grow or Die!* (New York, Arlington House, 1977).

149 Sandra Wenderoth & Edward Yost, *Multispectral Photography for Earth Resources* (New York, C.W. Post Center, Long Island University, 1974).

150 Elizabeth Whelan and Frederick Stare, *Panic in the Pantry* (New York, Atheneum, 1977).

151 Wing-tsit Chan, *A Source Book in Chinese Philosophy* (Princeton, N.J., Princeton Univ. Press, 1963).

152 J.Z. Young, *The Life of Mammals* (London, Oxford Univ. Press, 1967).

INDEX